HISTORY OF POLYOLEFINS

CHEMISTS AND CHEMISTRY

A series of books devoted to the examination of the history and development of chemistry from its early emergence as a separate discipline to the present day. The series will describe the personalities, processes, theoretical and technical advances which have shaped our current understanding of chemical science.

HISTORY OF POLYOLEFINS

CHEMISTS AND CHEMISTRY

A series of books devoted to the examination of the history and development of chemistry from its early emergence as a separate discipline to the present day. The series will describe the personalities, processes, theoretical and technical advances which have shaped our current understanding of chemical science.

HISTORY OF POLYOLEFINS

The World's Most Widely Used Polymers

Edited by

RAYMOND B. SEYMOUR
Department of Polymer Science,
The University of Southern Mississippi, U.S.A.

and

TAI CHENG
Raychem Corp., Menlo Park, California, U.S.A.

SPRINGER-SCIENCE+BUSINESS MEDIA, B.V.

Library of Congress Cataloging in Publication Data

CIP

Main entry under title:

History of polyolefins.

(Chemists and chemistry)
Includes bibliographies and index.
1. Polyolefins–History–Addresses, essays, lectures. I. Seymour, Raymond Benedict, 1912- . II. Cheng, Tai. III. Series.
TP1180.P67H57 1985 668.4'234 85-19632

ISBN 978-94-010-8916-6 ISBN 978-94-009-5472-4 (eBook)
DOI 10.1007/978-94-009-5472-4

TABLE OF CONTENTS

PREFACE

The word Polyethylene was probably first pronounced in a lecture which M. P. E. Berthelot delivered on April 27, 1863 to the Chemical Society in Paris, reporting on the "polymerization" of various simple organic compounds (1). Much later this work appeared twice in the literature before the classical ICI breakthrough in the 1930's which is so colorfully described in Ballard's lecture.

Once it came up at the end of the last century when H. von Pechmann obtained "a white flocculant material" from the decomposition of diazomethane which, one year later, was termed to be "polymethylene" - (CH_2) - from E. Bamberger and F. Tschiemer (1). At that time the investigators were disappointed about this product because it was not what they had expected to find in their experiments. As a result any further work was discontinued. The second time that the word polyethylene appeared in the literature to describe a "white solid powder" was in 1930 when C. S. Marvel and M. E. P. Friedericks (2) attempted to prepare alkylated As compounds in which all five valencies were covalently bonded to five monovalent alkyl groups. They reacted Tetra-ethyl-arsenium bromide with butyllithium and expected to get tetra ethylbutyl arsenium.

$$(AsEt_2) + Br^- + BuLi \qquad LiBr + AsEt_4Bu$$

Instead they obtained $LiBr + AsEt_3$ + gaseous products.

Delicate and somewhat time-consuming analysis gave a surprising result: ethane and C_4's were there in the expected quantities but ethylene was missing - or almost missing - in the gas mixture.

This again could have been the end of this study but Speed wanted to know exactly why it was not there and passed ethylene gas through a solution of n-BuLi in a high boiling mineral oil at somewhat elevated temperatures. He obtained a white, solid powder: linear polyethylene. Here was the first solid polyethylene in excellent yield by direct addition polymerization under very mild conditions with a metal organic catalyst: Twenty four years before Karl Ziegler's famous polymerization of ethylene with coordination complexed catalysts. Seven years had to pass before ICI chemists would develop the high pressure and high temperature synthesis of low density polyethylene-Polythene-and 20 years before the scientists and engineers of the Phillips Petroleum Company produced a high density species - Marlex - at intermediate pressures and temperatures. In 1930 polyethylene did not seem to be a very attractive material and the catalyst, n-butyllithium, was still a laboratory curiosity. The duPont Company, where Marvel had just started to become a technical consultant was busy with nylon, Neoprene, Lucite and several vinylpolymers and there was no immediate follow-up on this "curious" observation.

The word "polypropylen" was probably first printed in German in two articles by A. Butlerov which appeared in den Annalen der Chemie in 1876 and 1879.

At that time it was neither isolated nor described in any detail. It took a long time until the polymerization of propylene was again studied but now in great detail by C. M. Fontana and associates at the laboratories of the Socony Mobil Company in Paulsboro, New Jersey. At that time the interest focussed on the synthesis of polymers which could be used as efficient "viscosity index modifiers", i.e. compounds which, if added to luboils would flatten the viscosity-temperature curve (3). Fontana used aluminum bromide as catalyst and HBr as "promoter". Under relatively mild conditions he obtained high molecular weight polypropylene having a very broad molecular weight distribution (i.e. from about 600 to 830,000). Because of extensive hydryde transfer the molecules had a highly branced - Christmas tree-like conformation and showed no tendency to crystallize. However, the fact that propylene does polymerize with cationic (acid) catalysts, similar to isobutylene was a strong incentive to include it very early into the list of candidates for coordination complex catalysts.

H. Mark

REFERENCES

1. Quoted from Herbert Morawetz, "Polymers, The Origin and Growth of a Science," J. Wiley, New York, 1985.

M.P.E. Berthelot References 52 and 52b.

H. von Pechmann Reference 52d

E. Bamberger and F. Tschiemer Reference 52c.

A. Butlerov Reference 72.

2. C.S. Marvel and M.E.P. Friederich, JACS 52, 376 (1939).

3. C.M. Fontana, G.A. Kidder and R.J. Herold; Ind. Eng. Chem. 44, 1688 (1952).

C.M. Fontana, R. J. Herold, E.J. Kinney and R.C. Miller; Ind. Eng. Chem. 44, 2955 (1952).

C.M. Fontana; Ind. Eng. Chem. 63, 1167 (1959).

We would like to express our sincere appreciation to all the contributors for their patience, cooperation and efforts in preparing the material for the first history book of polyolefins. We, on behalf of the Division of History of Chemistry of the American Chemical Society, also would like to thank the following companies for their generous financial assistance. Without their support, this book would not be made possible for publication. The names of these companies are Allied Corporation; Amoco Chemicals Corporation; Chevron Research Company; E.I. du Pont; Exxon Chemical Company; The Firestone Tire & Rubber Company; Gulf Research & Development Company; Himont Incorporated; Eastman Kodak Company; Loctite Corporation; Mallinckrodt, Inc; Monsanto Company; Phillips Petroleum Company; The Procter & Gamble Company; Raychem Corporation; Stauffer Chemical Company; 3M Company; Union Carbide Corporation; and W.R. Grace & Company.

Tai Chun Cheng and
Raymond B. Seymour

Frank M. McMillan

Remarks at the Symposium on the History of Polyolefins, Miami Beach, FL, Apr. 29, 1985

Questions have been raised at this symposium regarding the accuracy of published accounts of discoveries in the polyolefin field, and the papers presented here have reflected markedly different viewpoints about the significance of some of these events. As the author of one such published account*, I should like to offer a comment on the problem of properly allocating credits and priorities.

There were many workers in the polyolefin vineyard, and many contemporary discoveries. Some of the pioneers have complained that their work did not receive as much recognition as it deserves. Some of these plaints have merit; however, the papers presented at this symposium dramatically illustrate another relevant point that should be of concern to the American Chemical Society and, in particular, to the Division of the History of Chemistry.

The reason that the names Ziegler and Natta are household words in the house of polymers is not only that they made key discoveries, but also that they were prompt and thorough in publishing their findings: early, detailed, and complete publication, not just in the fairyland of patent applications but in the scientific literature, and not 25 years after the fact.

In this way, their work received immediate attention, review, criticism, and appreciation by their peers.

I suggest that if the industrial laboratories in the United States (duPont, Phillips, Standard of Indiana, for example, perhaps also Exxon, Eastman, Hercules, Goodrich, and others) had not only permitted but actively encouraged their scientists to create research and publication records similar to those of Ziegler and Natta, their recognition would have been higher and less belated, and the work of the historian would have been made much easier.

* "The Chain Straighteners", Macmillan Publishers Ltd., London, 1981; ACS Books, Washington, DC.

Raymond B. Seymour

INTRODUCTION TO HISTORY OF POLYOLEFINS

Introduction

The history of polyolefins is usually considered to begin with the discovery of Fawcett and Gibson at ICI in England in the early 1930's. Nevertheless, the naturally occurring polyolefin, elaterite and related bitumins have been known for centuries. Of course, the homologous alkane series has been known for many decades but even the solid paraffin waxes were not considered to be polymers.

Synthetic polyolefins were synthesized by the decomposition of diazomethane by E. Hinderman (1), H. von Pechman (2), E. Bamberger and F. Tschirner (3) in the 1890's, by H. Meerwein and W. Burneleit (4) in 1928 and by W. Weile (5) in 1938. Polyethylidine and polypropylidine were also obtained by the catalytic decomposition of diazoethane and diazopropane, respectively by G. D. Buckley, L. H. Cross and N. H. Ray (6) in 1950.

Liquid oligomers of isobutylene were produced by H. A. Butlerov and B. Goryainov (7) in 1873 and higher molecular weight isobutylenes were made by the boron trifluoride cationic polymerization of isobutylene by M. Otto (8) in 1927. Polyolefins were also produced by H. Staudinger (9) in the early 1920's by the hydrogenation of natural rubber and by W. H. Carothers in 1930 by the Wurtz Condensation of decamethylene dibromide in 1930.

With the exception of polyisobutylene, which was commercialized by I. G. Farbenindustrie, these polyolefins were essentially laboratory curiosities. They could not be produced ecomonically and with the exception of H. Staudinger and W. H. Carothers, few recognized either the existence or the importance of macromolecules.

R. B. Seymour and T. Cheng (eds.), History of Polyolefins, 1–7.

LDPE

Hence, the invention of low density polyethylene (LDPE), as described by D. G. H. Ballard in a subsequent chapter, was not only accidental but propitious because of a recognition of polymer science concepts, the introduction of techniques for reactions at high temperatures, the cooperative investigations of ICI and Dr. Michels at the University of Leyden in Holland and the willingness of ICI to gamble on the possible conversion of a laboratory curiosity into a commercial product.

The discovery of LDPE, like that of many polymers, was accidental or serendipitious but chance favored those who were prepared. Nobel Laureate (1947), Robert Robinson proposed an investigation of the reaction of benzaldehyde and ethylene at high pressures. No condensation occurred but Eric Fawcett and Reginald Gibson, who performed these experiments observed the presence of less than 1 gram of a solid hydrocarbon.

Even Nobel Laureate (1953), Herman Staudinger maintained that ethylene would not polymerize and hence an offhand remark on the formation of LDPE by Fawcett at a Faraday Discussion in 1935 was not taken seriously by anyone except those executives at ICI who abruptly terminated Fawcett's career as a polymer scientist. It should be mentioned that Fawcett had spent over a year working with W. H. Carothers in his laboratory at duPont and was well aware of the existence of macromolecules.

ICI failed to patent LDPE. However, the company did file patents for the production of LDPE (then called high pressure polyethene, HPPE), in 1936, three years after Fawcett and Gibson had produced the first LDPE sample and one year after Fawcett had made his unauthorized statement at the Faraday Discussion.

In spite of these delays, LDPE was produced commercially in 1939 and its use as an insulator in coaxial cable was essential for Britain's World War II Radar system for spotting approaching enemy aircraft.

This and subsequent developments in polymer science led Lord Todd, President of the Royal Society of London, to state: "I am inclined to think that the development of polymerization is perhaps, the biggest thing chemistry has done, where it has had the biggest effect on everday life" (10). This statement applies equally well to LDPE.

HDPE

With the exception of the highly branched LDPE, all the other polyethylenes produced for the decomposition of diazomethane and the Wurtz condensation of alkylidene dichlorides were linear polymers, like the normal alkane homologues. Linear polyethylene, formerly called low pressure polyethylene (LPPE) and now called high density polyethylene (HDPE) was produced independently in the early 1940's by C. S. Marvel and F. Mayo but neither duPont or U.S. Rubber Company recognized the commercial importance of these higher melting polymers (11). Marvel and his student M. Friedrich used lithiumalkyl and an arsonium compound (12) as a catalyst for the polymerization of ethylene in the 1930's.

Also, in 1937, F. Hall and A. Nash observed the formation of aluminumalkyls when attempts were made to use aluminum chloride as a catalyst for the polymerization of ethylene. In 1943, M. Fischer of Badische Aniline und Sodafabrik used aluminum powder and titanium tetrafluoride as a catalyst for the polymerization of ethylene. A patent assigned to BASF was issued in 1953.

The commercialization of HDPE was made possible, in 1953, when Karl Ziegler (1963) discovered the "Aufbau" reaction for producing linear polyethylene. Again, serendipity played a part. The first product was a dimer (1-butene) because of contanination by nickel from the the reaction vessel which terminated the propagation.

This discovery by Ziegler's graduate student C. Holzkamp led to a systematic search for other metals which hopefully, would enhance the propagation reaction. The best system was found to be titanium tetrachloride and aluminumalkyl (13). Reverse serendipity also played its role when Dr. Ziegler wrote his own patent application and confined his claim to the polymerization of ehtylene (14).

As discussed in a subsequent chapter by P. Hogan and

R. Banks, G. Baily and E. Reid of Phillips Petroleum Company polymerized ethylene using a nickel oxide catalyst on an alumina-silica supportin 1945. Later, Hogan and Banks made HDPE by adding a chromium salt to the Baily-Reid catalyst system and were granted a patent for this process in 1953. Also, A. Zletz of Standard Oil of Indiana (now called Amoco) applied for and was granted a patent in 1951 for the production of HDPE using a molybdenum oxide catalyst supported on alumina.

POLYPROPYLENE

The "Ziegler Catalyst", titanium chloride and triethyl-aluminum was licensed to several companies, including Montecatini, which is now called Montedison. In 1954, P. Chini, working in the laboratory of Nobel Laureate Natta (1963), used this catalyst to produce polypropylene. Accordingly, Natta wrote in his notebook, "today (March 11, 1954), we made polypropylene" (15). Natta was consultant for the Montecatini Company.

As evident by discussions in subsequent chapters, chemists in many of the laboratories of Ziegler licensees also produced polypropylene. However, while a patent was granted by the U.S. Patent Office to G. Natta in 1973, the U.S. Federal District Court ruled that this patent was invalid and concluded that Hogan and Banks had invented crystalline polypropylene in 1951.

COPOLYMERS OF PROPYLENE

While Montedison spent many years and several million lire in its unsuccessful attempt to prove that Nobel Laureate Natta (1963) was the inventor of polypropylene, this company failed to emphasize Natta's most important contribution i.e., the elucidation of the concept of stereo - specificity which he called tacticity. Polypropylene, with a random sequence of l and d repeating units is a soft amorphous polymer. In contrast, commercial polypropylene contains at least 95% of isotactic (dddd or llll) polypropylene. This highly ordered polymer is crystalline and has a relatively high melting point.

As related in a subsequent chapter, by E. Tornquist, G. Bier, one of Ziegler's assistants, produced amorphous elastomeric copolymers of ethylene and propylene in 1954 (16) and Natta and coworkers showed that vanadium-tetrachloride was a better catalyst than titanium tetrachloride for this copolymerization. These elastomers were commercialized by EXXON in 1960 and duPont produced a vulcanizable elastomer (EPDM) in 1963 by using a small amount of norbornene as a comonomer in the ethylene-propylene system.

BUTYL RUBBER

Like the discovery of EPDM, that of butyl rubber by R. Thomas and W. Sparks was based on a knowledge of polymer science. As discussed in a subsequent chapter by E. Tornquist, both IG Farbenindustrie and Standard Oil of New Jersey (now called EXXON), produced polyisobutylene in the 1930's, using information developed by M. Otto (8). In 1937, Thomas and Sparks used aluminum chloride to polymerize a mixture of isobutylene and isoprene and this vulcanizable elastomer was called butyl rubber.

Since it retained air better than natural rubber, butyl rubber was used for inner tubes. This resistance to gaseous permeation was improved dramatically by chlorination and bromination, in which the halogens were substituted for hydrogen atoms and did not reduce the unsaturation of the vulcanizable elastomers.

LLPDE

Copolymers of ethylene and 1-butene have been produced commercially in Japan for over two decades and duPont has operated a Canadian plant since 1960 which uses a solution process to produce polyolefins. As discussed in a subsequent chapter by F. J. Karol, Phillips Petroleum Company announced a slurry process for the copolymerization of olefins in 1969 but most of its copolymers had specific gravities greater than 0.93. Another slurry process and a gas phase process for producing lower density linear copolymers of ethylene (LLDPE) were announced by Dow and Union Carbide, respectively in the early 1970's.

Because of prior investment in production facilities, large quantities of LDPE continue to be produced. However, had LLDPE been available in the 1903's, it is doubtful that the higher pressure process for producing LDPE would have been commercialized.

CATALYSTS

After the accidental polymerization of ethylene at extremely high pressures by Fawcett and Gibson, it was discovered that a trace of oxygen had served as the catalyst for this unanticipated "break through" in polymer science. The discovery of the more sophisticated "Ziegler Natta" catalyst system was another important "break through."

However, the most important of these developments in catalysis was the development of a unique system for the improved polymerization of various olefins in the 1980's. Some of these new developments which have been called a "catalyst revolution" are discussed in subsequent chapters. It may be anticipated that this revolution will continue. However, it is also important to record the events leading up to these revolutions. Fortunately, much of this history is being recorded while many of the pioneers who advanced polyolefin technology are still alive.

REFERENCES

1. Hinderman, E., Dissertation, Zurich, 1897.
2. von Pechman, H., Ber 31, 2643 (1898).
3. Bamberger, E., and Tschirner, F., Ber 33, 955 (1900).
4. Meerwein, H., and Burneleit, W., Ber 618, 1845 (1928).
5. Weile, W., Dissertation, Marlburg, 1938.
6. Buckley, G. D., Cross, L. H., and Ray, N. H., J. Chem. Soc., P2714 (1950).
7. Butlerov, H. A., and Goryainov, B., Ber 6, 561 (1873).
8. Otto, M., Brennst-Chem., 8, 321 (1927).

9. Staudinger, H. Ber 57, 1203 (1925).

10. Chemical and Engineering News, 58, 29 (1980).

11. McMillan, F. M. "The Chain Straightners," McMillan Press LTD, London, 1979.

12. Fredrich, M. E. P., and Marvel C. S., J. Am. Chem. Soc., 52, 376 (1930).

13. Ziegler, K., and Martin, H., Angew Chem, 67 (19/20) 541 (1955).

14. Seymour, R. B., Pop Plastics, 30 (1) 16d (1985).

15. Seymour, R. B., Pop Plastics, 30 (1) 16f (1985).

16. Bier, G., Angew Chem, 73, 186 (1961).

D. G. H. Ballard

THE DISCOVERY OF POLYETHYLENE AND ITS EFFECT ON THE EVOLUTION OF POLYMER SCIENCE

Introduction

The juxta position between the first and second World Wars was responsible for a major increase in all forms of technology. Associated with this was the creation of the novel materials necessary for these new technologies to become possible. What is staggering in retrospect was the rate at which these new materials were created and brought into use. Polyethylene, for example, although first discovered in 1933 was not a reproducible technology until 1935 but by the outbreak of the second World War in 1939 a production a production plant was in operation and a telephone cable between the Isle of Wight and Mainland England was in use. Similarly with ICI's other development Perspex (Polymethylmethacrylate) and Du Ponts invention of Nylon, all of which occurred about the same time as Polyethylene, were also available within four to five years.

ICI, or Imperial Chemical Industries to give the full name was a new company formed in 1927 by the amalgamation of British Dyestuffs Company Brunner Mond, United Alkali and Nobel Explosives Industries (1). It was formed for strategic reasons to create an organisation to compete with the German chemical giant, I.G. Farben. The glue that welded these companies together was the large sums of money available from the explosives company for research and the protection offered by the British Government which virtually excluded the European Chemical Industry from the, whole of the British Empire and resulted in ICI becoming an international company. To further strengthen its position it formed an alliance with the Du Pont family which was unique in that the research communities of both organisations behaved as one and there was a free exchange of information. In addition to the industries which these two great organisations created they produced some of the most original polymer scientists of the time.

R. B. Seymour and T. Cheng (eds.), History of Polyolefins, 9–53.

The Origin of the High Pressure Technology

In the 1930's chemical kinetic's was an evolving subject and several of the British Schools were involved in studies of the kinetics of organic reactions, this being the main thrust in comprehending the mechanisms of the formation of organic compounds. At the same time engineering techniques were being made available which enabled very high pressures to be achieved, very high temperatures and also very hard vacuums. A programme of work on basic chemistry at these extreme conditions had been conceived by F A Freeth for the new laboratories of Alkali Division of ICI at Winnington in Cheshire which were completed in 1928. He also recruited a talented group of chemists, physists and engineers. Also Freeth had connections with the Thermodyanmics Laboratory at the University of Leyden, Holland and especially with Dr A Michels. The latter became a consultant to ICI and advised on the design of an apparatus capable of working at pressures up to 12,000 atmospheres.

F A Freeth was a creative research manager determined that good basic science should be done at the Winnington Laboratory. He spent a considerable amount of his time visiting University Research departments and meeting potential recruits. He was directly responsible for recruiting Michael Perrin and R O Gibson and for placing them in Michels laboratory. Perrin was there from 1929 to 1933.

The programme of study which lead directly to the discovery of polyethylene was conceived in two ICI research centres independently. The first proposal was written by Michael Perrin and J C Swallow in 1931, while Perrin was in Michels Laboratory. At about the same time Sir Robert Robinson, who was an adviser to the research department at Dyestuffs Division, invited Michels to give a lecture at Blackley on techniques for generating very high pressures. As a consequence Sir Robert Robinson proposed(1) that certain organic reactions should be studied to see if they would go without catalysts. Included in this list was the reaction between ethylene and benzaldehyde. The consequence of this was that E W Fawcett, an organic chemist went to work at at Winnington with R O Gibson to combine the skills of an organic chemist with high pressure technology.

At the time the business interests of Alkali and Dyestuffs Divisions did not include polymeric materials. Moreover it is a matter of record(2) that the work at Winnington was exploratory chemistry in an area where no theoretical background existed. In all Fawcett and Gibson studied some 50 reactions. They made a significant contribution to chemistry in that they identified clearly what was to be expected of applying high pressures to organic reactions. Their main conclusion was that reactions which proceeded slowly at atmospheric pressure had their rates increased 5 to 10 times at pressures approaching 3000 atmospheres.(2,3) Reactions which did not occur at atmospheric pressure could not be induced to take place by the application of these very high pressures.

In the early 1930's polymerisation chemistry was primitative. Several of the vinyl monomers had been polymerised but the roll of free radical chemistry was not appreciated and the use of organic compounds, which could produce free radicals by thermal decomposition, were not used. The organic polymer chemist had arrived as certain empirical rules concerning the polymerisation of olefinically unsaturated vinyl compounds. One of these rules was that ethylene was too inactive a molecule to be induced to polymerise. This view persisted among academic scientists to the point of dogma, for when Fawcett reported in 1935, at the Faraday Discussion on Polymer Chemistry, that ICI had succeeded in polymerising ethylene it was dismissed by Staudinger as nonsense(4).

It was not until the 29th March 1933 that Eric Fawcett and Reginald Gibson produced the first fraction of a gramme polyethylene but subsequent experiments were not reproducable and decompositions occurred leading to explosions. The instability of the system lead the sponsors of the programme, Dyestuffs Division, to withdraw their support and Fawcett was transferred to other work. The fact that the early experiments were carried out in open laboratories made them extremely hazardous. However, by the time W H Perrin had taken charge of the programme in the spring of 1935, W R D Manning a Research Engineer, had developed equipment able to handle gases at these high pressures with safety.

It is highly unlkely that polyethylene would have been an important war time material for radar but for Michael Perrin. The latter joined the Northwich group in 1935 from Michels laboratory and changed the whole approach. His concern was to study mechanisms of reactions and was greatly influenced and advised by Professor Hinshelwood of Oxford. The original experiments of Fawcett and Gibson were concerned with the reaction of benzaldehyde and ethylene. What is now known is that oxygen contamination of the ethylene was important because this formed peroxides which decomposed to give a free radical initiator. Without a catalyst, ethylene will not polymerize spontaneously under pressure and if heated above 300°C it will more than likely decompose into carbon, methane and hydrogen. Also the benzaldehyde of the day probably also contained peroxide impurities. Knowledge of the relationships between catalyst concentration and rates of polymerization were beginning to be studied in the mid-thirties and Perrin was no doubt aware of this from his association with Hinshelwood. In any event, Perrin's systematic approach was immediately successful. On the 19th of December, 1935, he removed benzaldehyde from the reaction mixture of Fawcett and Gibson and studied the behaviour of ethylene under a pressure of 2,000 atmospheres. Instead of trivial quantities of the polymer being obtained, the reactor this time lost pressure in a smooth and controlled way as ethylene polymerized. Repeated additions of ethylene led eventually to the first batch of 8 grams of good polymer being produced.

The main problem of identifying who did what and when is that some of the reports are inaccurate due no doubt to faulty memories, since most accounts are post 1944. For example, M W Perrin, who took charge of the experimental programme in 1935, was reporting to J C Swallow, the Laboratory manager at the time. In one of J C Swallow's accounts of the history (5), M W Perrin is mentioned as suggesting that high pressure experiments be carried out. This is in 1932, a year later than the official date of the proposal. Perrin in 1932, although an ICI employee was studying at Amsterdam University with Professor Michels. Swallow omits the fact that in the two accounts written by him that Perrin actually did the critical experiment. Perrin, Gibson and Fawcett had all left ICI by the end of the war and were not involved in subsequent developments. The ICI official historian (1), however, seems to have gone into the matter with great care. Also a recent article in the New Scientist is concise and accurate (New Scientist March 24, 1983 p 836). Finally M W Perrin became Sir Michael Perrin and achieved great distinction as a Research Director during the war concerned with the early phases of the atom bomb, later Chairman of the Wellcome Foundation, etc, etc. At the time of writing he is alive and although eighty years of age still very active.

The contribution that M W Perrin made to the discovery of polyethylene was crucial. He succeeded in defining reaction conditions which enabled polyethylene to be synthesised reproducably. Fawcett and Gibson, when they encountered explosions and unreliable behaviour, lacked the physical chemist's approach of Perrin which enabled the problem to be solved. Reduction of the process to practice demanded reproducability. Without this ICI could not have developed the technology in time for manufacture during the 2nd World War. What occurred subsequently was a unique piece of process development in which a continuous process for the manufacture of polyethylene at 1,500 atmospheres evolved.

Following the second war ICI licensees DuPont and Union Carbide both made substantial contributions in finding a wide range of applications for polyethylene which eventually lead to massive increases in the scale of manufacture.

It is worthwhile mentioning some of the features of this research community since the young men who were involved subsequently became very senior managers and contributed to the building of ICI into one of the worlds larger chemical companies. The predominont discipline was physical chemistry and the majority had been recruited from Oxford University. The laboratories were new and located close to a Cheshire "Black and White" country house (built in 1720) which was the cultural centre. Leasure persuits resembled those to be found in an Oxford college, namely good food, mature wine, and bridge and snooker in the evenings. In the summertime croquet and tennis could also be played on the lawns outside. All of this against a backcloth of a 50 year old Alkali Works in the middle of Cheshire, one of Englands very rural counties. To be a member of Winnington Hall Club was a privilaged reserved for chemical gentlemen and senior managers. In accordance with the English order of professional merit of the time Engineers could not possibly be gentlemen and were therefore not allowed to be members of the Club. There was of course no women and wives of members were not allowed at the bar and had to enter the club via the french windows. In 1960, when ICI had grown to and diversified into ten divisions each of which had a research director, all of them were Oxford graduates. In the present organisation there are no Oxford research directors but Winnington Hall Club is still there and women are permitted. Finally in the community of early 30's it was considered that the PhD was an inferior German invention and holders of such a qualification were always addressed as Mister. It should be pointed out however that the availability of people trained to a level of PhD was very limited. This was because research schools in the principle Universities did not have the funds to enable even their best students to stay for three years, the minimum required for a doctorate. Consequently the majority of scientists entering industry at the time were trained to the level of a batchelors degree which, if you wished to be regarded as a physicist or chemist required three years study. The exception was Oxford which had a fourth year in which students were given basic training in the Oxford research schools and allowed to add a BSc to their BA degree. The two important groups in chemistry at Oxford were the organics school of Robert Robinson and the physical chemistry school of Cyril Huchalwood both of whom became Nobel Laureates. Also they both acted as consultants to the ICI research groups at Winnington and Blackley. At Manchester University which was about 20 miles away two of the founders of the Transition State Theory of Chemical Kinetics were in residence namely M.G. Evans and Michael Polanyi.

The group that created the high pressure manufacturing process was lead by E Hunter whose main feat was the development of a stirred reactor operating at 1500 atoms. The problem of creating a pressure secure gland for the stirrer shaft was eventually solved by not having one. The electric motor which was above the shaft was put into the reactor and the wires carrying the current were insulated from the reactor wall by glass-metal seals. This design has been scaled to the level of several hundred thousand tons of polyethylene per annum on single stream units and operated in most countries in the world. The scale off world production is now between 5 to 10 lbs per head of population.

Polyethylene and the Evolution of Polymer Science

(a) Polymerisation Chemistry and the Structure of Copolymers

It is difficult, in retrospect to visualise the difficulties of studying macromolecules in the period between the wars. Physical methods for characterising polymers chemically and for measuring molecular weights were minimal and made more complex by the presence of functional groups. The advantage that polyethylene had was the very simplicity of its chemical structure, its crystallinity and the fact that its importance resulted in a vigorous programme of study on both sides of the Atlantic. Also as was subsequently found ethylene can be polymerised using radicals, anionic and cationic initiators as well as the wide variety of organometallic catalysts which are now available. This allowed the chemistry of polymerisation to be studied in a way not possible with more complex monomers. This chemistry is to be discussed by other contributors to the symposium. The high pressure polyethylene produced by radical polymerisation was to present some serious problems to those seeking to make the polymers at low pressures. The main reasons for this have only recently been fully appreciated.

To begin with the remarkable effect of pressure on the polymerisation of ethylene was explained by Evans and Polanyi (6) in 1936 who applied their transition state theory of reaction kinetics to the problem and obtained the equation:

$$\frac{d[\ln k]}{dp} = \frac{\Delta V^*}{RT} \qquad (1)$$

$$\Delta V^* \simeq V_m - V_p$$

where k is the velocity constant for the reaction, p the hydrostatic pressure, and $\Delta V^{\#}$ the difference in volume between the solution of reactants and the complex in the transition state. It is considered that the volume of the

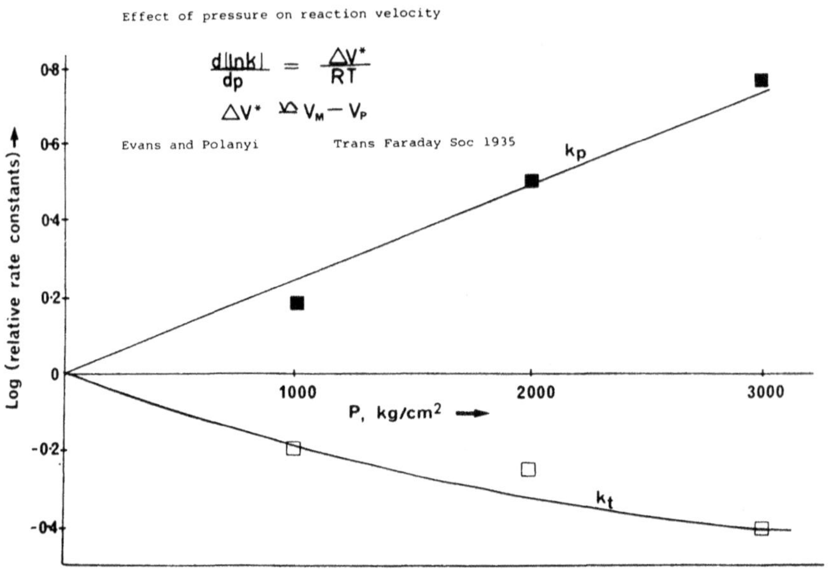

Effect of pressure on k_p and k_t, in radical polymerisation of Styrene at 30°C

Figure 1

latter is intermediate between reactants and products but closer to the latter. Of the reactions, which determine molecular weight, namely initiation, propagation and chain termination by all processes, the growth step is the most affected by pressure because of the contraction in volume on the formation of polymer. Measurements carried out subsequently on a variety of polymer systems has shown that the velocity constant for propagation, kp, is markedly increased by hydrostatic pressure (7). This is shown in figure 1.

Important new information has recently become available on the relationship between the structure of the comonomer and its ability to control structural parameters on a scale of 50-150°A. This information derives fom two converging studies. The first is the improvement in our knowledge of the conformation of the macromolecule in the solid state derived from small angle neutron scattering (SANS) and small angle X-ray scattering (SAXS). The second is the availability of highly active organometallic catalysts which give a more precise control of the microstructure of the copolymer.

The mechanism of formation of short side of chains in chains in radical polymerisation is as follows:-

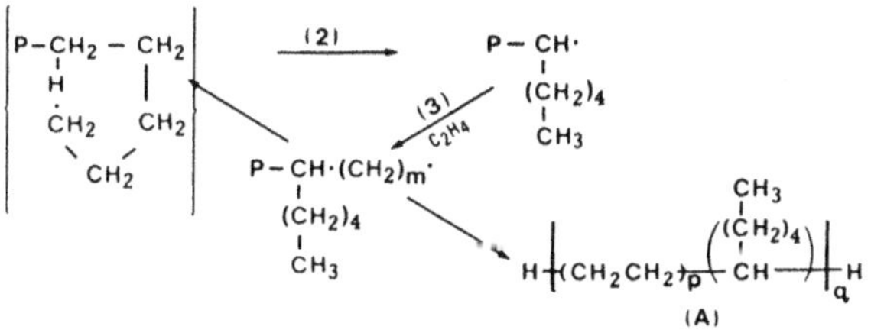

The number of branches are statistically distributed and determined by the relative probability of propagation and transfer reactions (7).

Table 1

MEASUREMENT OF THE PARTITION FUNCTION OF DISTRIBUTION OF -CHCl- units IN CHLORINATED POLYETHYLENE BETWEEN CHRYSTALLINE AND AMORPHOUS PHASES.

Mol Fraction of -CHCl- units per cent	Average distance between Chlorine Atoms A°	Crystallinity per cent	d A°	d_c A°	y
1·0	127	61 ± 4	188	115	1·06 ± 0·1
1·15	111	64	184	118	0·96
1·5	85	67	182	123	0·88
3·2	40	59	154	94	0·47

When y = 1 all CHCl units in the amorphous phase.

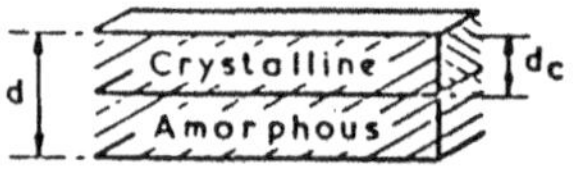

Introduction of short side chains using organometallic catalysis gave a greater variety of options because the length of the side chain could be varied.

$$C_2H_4 + R{\cdot}CH{=}CH_2 \longrightarrow \left[(CH_2{\cdot}CH_2)_n{\cdot}CH{\cdot}\overset{R}{\overset{|}{C}}H\right]_p \qquad (4)$$

Only recently developed heterogeneous catalysts can incorporate the comonomer uniformly. Previously polymerisations had been carried out above the melting point and a homogeneous catalyst used. Initially the procedures were empirical and the effect of the comonomer on the solid state structure was obscure. There was also confusion between the latter effect and those resulting from deliberate attempts to broaden the molecular weight distribution in HD polyethylene to improve processability. Much of this confusion derived from the fact that we were unable to study the organisation of the macromolecule in the solid state. Charles Bunn's X-ray study of polyethylene (8) was one of the more important steps in our understanding of polymer structure. At a time when there was still considerable debate about the character of macromolecules, the demonstration that wholly synthetic and crystalline polyethylene was a simple close packed strucure in which the bond angles and bond lengths were identical to those found in small molecules such as $C_{36}H_{74}$ and $C_{60}H_{122}$ (9), underpinned the strictly logical view that macromolecules were a multipliction of smaller elements joined by covalent bonds.

In figure (2) is shown the structure of polyethylene. Later Bunn predicted the correct crystal structure of polypropylene and higher polyolefins even though these could not then be synthesised. The unsolved aspects of the structure remaining were associated with the conformation of the macromolecule in the solid state. The two important models for polyethene are the fringed micell (10) (figure 3) and that based on the structure (supposedly known) of the single crystal (11) (figure 4). Small angle neutron scattering gave decisive support for

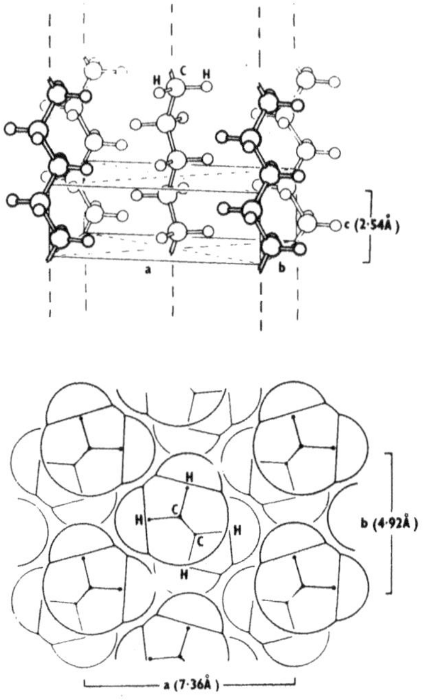

Figure 2

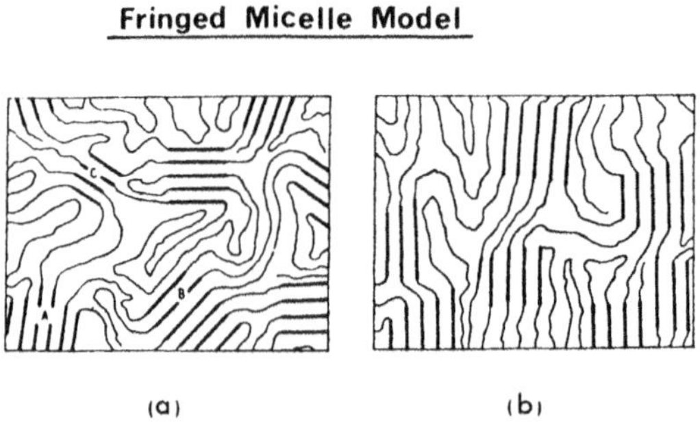

Schematic representation of (a) unoriented and (b) stretched, oriented crystallites

Figure 3

the view that the molecule was distributed in space and that the crystalline regions consisted of fragments of many thousands of different macromolecules.

The classical relationship between radius of gyration of a macromolecule and its molecular weight, previously only available from solution studies, could now be obtained from measurements in the solid state (12). Moreover measurements at intermediate angles coupled with computer simulation of competing models supported the lower angle work (13).

In figure 5 and 6 are shown the types of conformations which are consistent with recent structural studies. A summary of the work together with a discussion of the interpretation of the new data is given in the 1978 Faraday Discussion (12, 13), "The Structure of Macromolecules in Condensed Phases". Using these as a basis, we can now begin to analyse the role of the comonomer molecule in determining the properties of semi-crystalline polymers.

Empirically, it has long been recognised that the effect of copolymerisation with propylene was different from the effects observed if higher olefines were used. In particular the effectiveness of these two comonomers to reduce the density differed markedly. In figure 7 is shown the effect of two types of monomer on the annealed density of the copolymer obtained (15). The difference in behaviour was assumed to be due to the fact that a methyl side chain could enter the polyethylene crystal lattice but not the $CH_3(CH_2)_n$ group. The first direct evidence that this is so derives from a study of partially chlorinated polyethylene in which the chlorine atoms are randomly distributed. Using the differences in scattering cross section between the chlorine and CH_2 groups, Fisher and Schelten (16) were able to determine the distribution of chlorine atoms between the crystalline and amorphous phases at different comonomer concentrations. The partition function (Y) is given in table 1 for four copolymers. It is interesting to note that as the distance between chlorine atoms is reduced this is not matched by a proportionate reduction in d-spacing or the thickness of the crystalline lamellae. The values of (Y) show that above three <u>percent</u> chlorine approximately half the chlorine atoms are contained within the polyethylene crystal lattice. Figure 8a illustrates the way this occurs, the black dots representing the

Figure 4

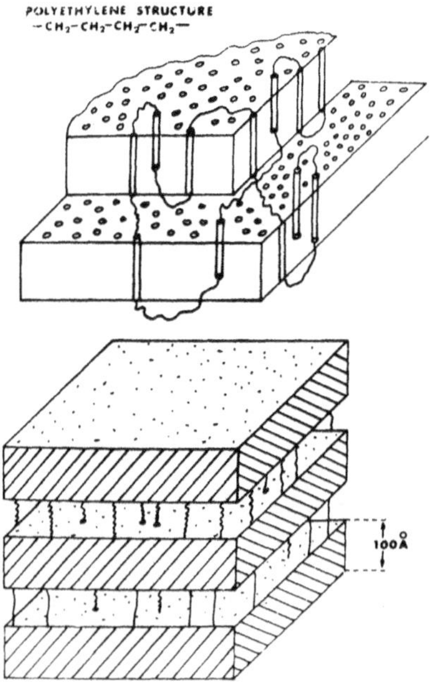

Figure 5

comonomer units. The methyl group in propylene is similar in diameter to CH_2 and behaves similarly and therefore has no specific function in controlling the proportion of the two dominant phases. Experiments of the latter type require a statistical distribution of comonomer units and this should be the same for all molecules independent of their lengths. This is described by the data given in figure 9a which is necessary to characterise copolymers. Figure 9a is an ideal copolymer in which the weight fraction of copolymer units is independent of molecular weight. The early types of heterogeneous Ziegler catalysts produced comonomer distribution curves of which figure 9b is typical. These polymers contain more comonomer in the low molecular weight molecules and such materials have poor properties.

It is possible to produce an ideal copolymer indirectly by hydrogenating polybutadiene. A polybutadiene produced in solution by anionic initiation contain a mixture of 1:4 and 1:2 units. It is possible to control the proportions by choice of initiator and polymerisation conditions.

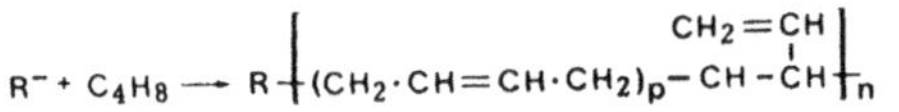
(5)

These polymers are completely soluble and hydrogenate smoothly to give a polymer which is structurally equivalent to a butene copolymer.

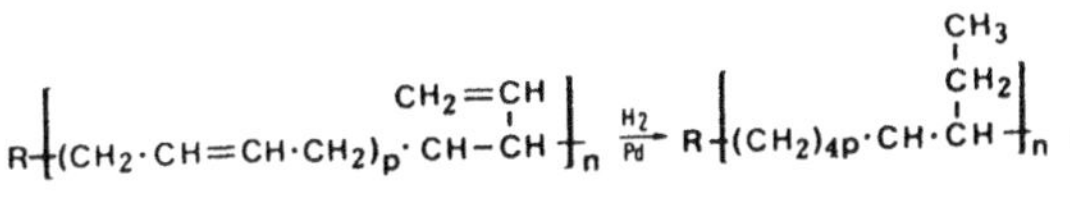
(6)

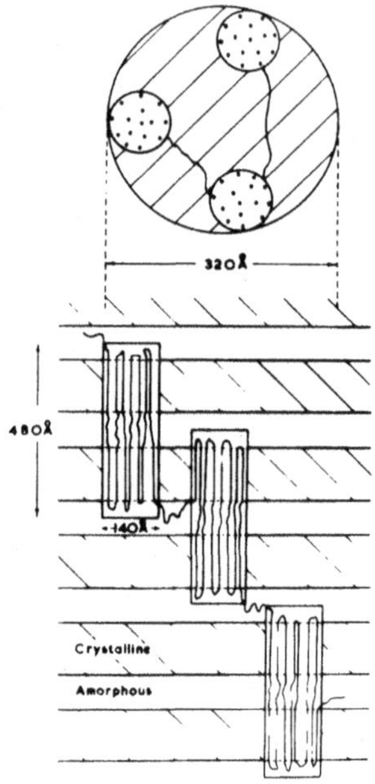

Figure 6

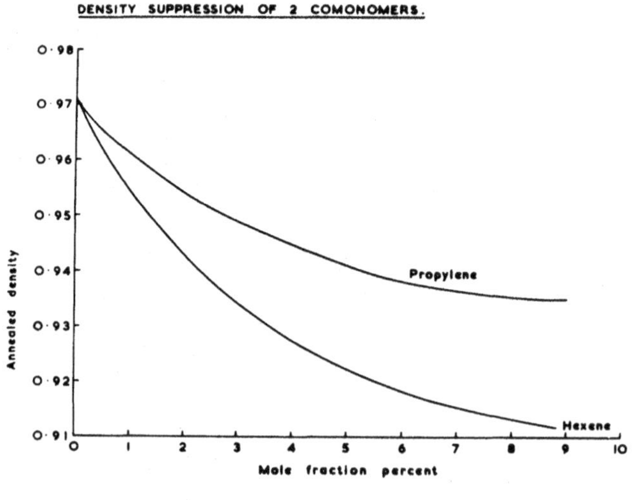

Figure 7

Such polymers are monodispersed and it is possible to study the effect of the comonomer unit, independent of the complexities of the molecular weight distribution.

In table 2 is summarised some of the feature of these polymers in which the fraction of branches determined by infra-red spectroscopy is 18-ethyl groups/1000C. This gives the average distance between comonomer units of 70A°. These copolymers do not change their crystallinity or d-spacing on annealing the comonomer unit effectively acting as a stop which prevents the crystallite thickening as shown in figure 9a. The crystal can grow to a thickness no greater than the distance between comonomer units. This seems to be approximately correct from the dimension given in table 2.

The thickness of the crystal is 50-55A° and the distance between comonomer units 70A°, the additional thickness is probably associated with the interface region between the amorphous and crystalline phases. The former, while not being crystalline, possess some organisation which inhibits the penetration by the comonomer units. It is therefore seen that this very precise copolymer has the characteristics of producing a two phase system in which the distance between comonomer units define the dimension of both phases as shown in figure 9a. It is also understandable why heterogeneous mixtures of copolymers have such poor properties, especially at low molecular weights.

If molecules of type B (figure 9b), are the predominant species, the crystallisation processes will generate a matrix with dimensions determined by the distance between comonomer units in these molecules. Molecules of type A, which have a higher concentration of comonomer units and hence are much closer together, will be rejected by the crystalline phase because it cannot accomodate the comonomer fragment. These molecules do not crystallise and inhabit the amorphous regions. Such materials are characterised by having polymers which can be extracted with solvents below the melting point.

A further interesting feature is that provided the comonomer units are uniformly distributed along the polymer chain, long chain branching does not affect the solid state properties. This is shown by polymer 2 in table 2 which was made via butadiene as before by using a trifunctional anionic initiator which gives a 3-arm star molecule. Each arm has a molecular weight of 33,000 and

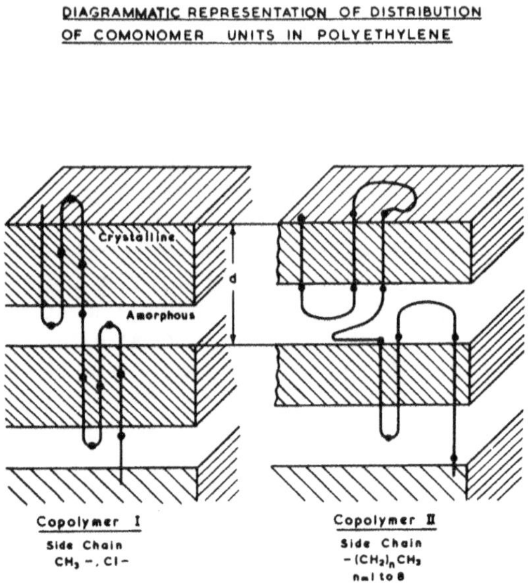

Figure 8

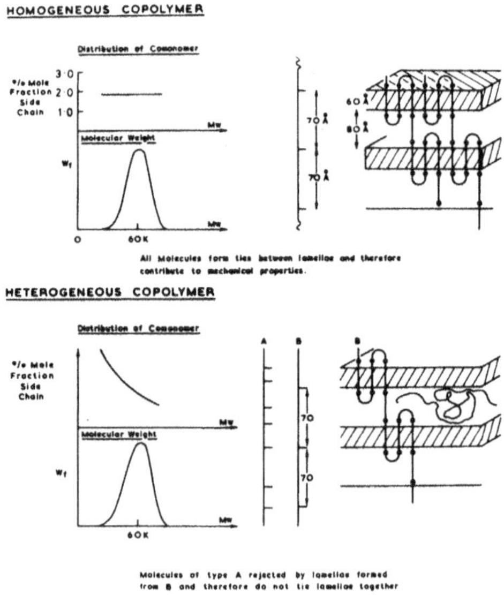

Figure 9

almost the same number of CH_2CH_3 groups per chain as for the linear polymer. It is evident from table 2, in this instance, that the solid state properties are unaffected by the geometry of the molecule. The only measurable difference related to the geometry is that radii of gyration measurements show that in the 3-arm star the radius of gyration is approximately 10 percent greater at the same molecular weight (17). This demonstrates that long chain branching, which is a feature of polyethylene produced by free radical polymerisation, is not important in affecting solid state properties. On the other hand where flow properties are concerned LLDPE is markedly superior to LDPE because this type of branching is absent.

It is evident that even the most successful organometallic catalysts polymerising under heterogeneous conditions do not produce a uniform distribution of comonomer units in all molecules, independent of molecular weight. This is shown by the effect of annealing and by the large d-spacing of polymer 3 in table 2. A similar situation applies to high pressure polyethylene. This may seem surprising since superficially the polymerisation is homogeneous. This is only partially correct since under the conditions of commercial manufacture, the ethylene is above the triple point and the system is multiphase with the degree of branching inevitably different in different phases and at the top and bottom of the reactor.

Organometallic catalyst even when operating under conditions where all the polyethylene is dissolved in a hydrocarbon solvent does not necessarily give a uniform distribution of comonomer among all macromolecules of differing molecular weight. This is because ethylene and the comonomer have different abilities to coordinate to the transition metal centre. The "polyethylene" synthesized via butadiene polymerisation and hydrogenation on the other hand is unique since the polymerisation only involves one monomer. The mechanism of insertion into the propagating polymer chain is by the same monomer predominantly adding 1:4 but occasionally 1:2.

In the case of heterogeneous catalysts working below the solubility of polyethylene the polymers produced have relatively broad distribution of molecular weight as well as a non-uniform distribution of comonomer units. It has been triumph of catalysis that systems can now be produced

Table 2

Polymers and Copolymers of the Polyethylene Type

Origins of Polymer	Mw	Mw/Mn	Tm °C	Side Chains CH_3's/1000	Distance between Side Chains Å	Crystallinity per cent		ρ gms/cc	Annealed	
						Quenched	Annealed		d(Å)	d_c(Å)
1. Homogeneous linear type copolymer synthesized via polybutadiene with CH_3CH_2- side chains	62,000	1	103	18	70	31	37	0.91	140	52
2. Homogeneous 3-arm star molecule with CH_3-CH_2-side chains on each arm Mol. wt. of each arm 33,000	99,000	1	103	16	79	35	37	0.92	145	54
3. Linear low density polyethylene with CH_3-CH_2- side chains.	98,000	4.33	118	19	--	41	50	0.92	170	85
4. High pressure polyethylene produced with radical initiators	125,000	14	108	22	--	39	52	0.92	135	70
5. Homopolymer made with similar catalyst to LLDPE	100,000	4.5	137	--	--	52	82	0.97	350	180
6. Copolymerisation of diazomethane with diazohexane. Ref. 23 The First LLDPE	70,000	--	115	30	--	38	--	0.92	--	--

which although operating heterogeneously in diluent or in the gas phase produce polymers with properties approaching those obtain in solution. One type of LLDPE produced in this way is shown in table 2.

Finally polymers produced by the copolymerisation of diazomethane with diazohexane closely resemble LLDPE. These were first reported in 1952 (18) but were actually made in the late 1940's.

(b) Inorganic Analogues of Polyethylene

The evolution of the science of organic macromolecules followed directly from the knowledge developed for simple organic molelcules. The valency rules for combining carbon atoms with other carbon atoms and with oxygen and nitrogen are the same for a macromolecule as for a simple compound. For example, it is easy for a graduate chemist to visualise the relationship between ethane, dodecane and polyethylene as primarily a matter of molecular weight thus:

CH_3CH_3 $\quad$ $CH_3[CH_2]_{10}.CH_3$ $\quad$ $CH_3[CH_2]_n\text{-}CH_3$

This simple idea combined with a system of converting any organic reaction which links molecules together into a polymerisation process has produced several homologous series of organic macromolecules.

No such simple rules exist for the synthesis of inorganic macromolecules. It is possible to synthesize linear inorganic macromolecules but no general rules are known. Frequently the route to these is by ring opening a cyclic monomer as in the case of sulphur, selenium and telurium. The products obtained are usually thermally unstable because of reversion to the ring form and offer little or no prospect for high temperature uses. The bulk of the studies have been on molecules which incorporate a significant fraction of organic atoms or groups. These organo-metalloid or organo-silicon polymers can be synthesized but only the silicons have advantages over organic materials for high temperature use.

In seeking new ideas for synthesizing completely inorganic materials it is better not to follow structural concepts originating from classical organic chemistry at the molecular level but to create analogues of the supra molecular structure observed in crystalline organic macromolecules in the solid state as shown in figures 5 and 6.

Recently (21) we have described work in which attempts have been made to produce analogues of

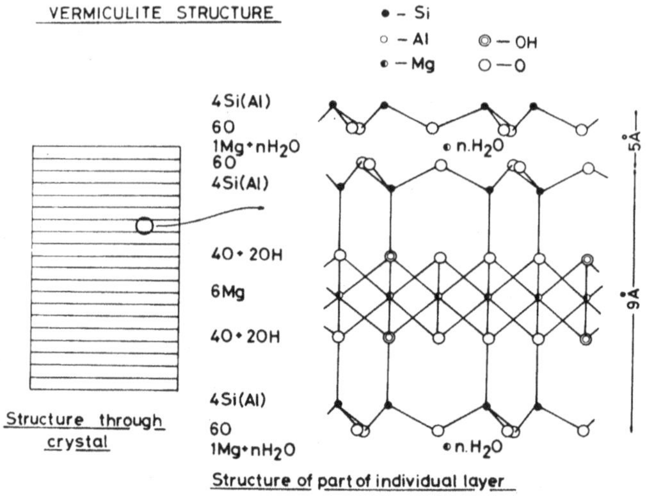

Figure 10

Figure 11

polyethylene and polypropylene by replacing the organic crystallites in structures shown in figure 5 with inorganic lamellae and linking these together using inorganic and organic tie molecules.

A readily available source of inorganic lamellae is to be found among a very large group of natural minerals known as the phyllosilicates including vermiculite, montomorillonite, hydrobiotite and chlorite-vermiculite. These have been studied extensively by scientists concerned with the structure of soils and the manufacture of ceramic materials (19). In their many uses advantage is taken of the very high surface area which enables them to interact with organic liquids as well as water.

The first attempt to produce flexible structures from layer minerals was due to G F Walker (20) who succeeded in making films from chemically delaminated vermiculite. Such films were of interest at the time but compared unfavourably with films made from organic materials such as polyethylene and polyethylene terephthalate (PET). At the time, Walker's work was carried out there was little evidence that materials formed by the assembly of lamellae could give structures which were flexible, strong and able to undergo moderate deformation. The fact that polycrystalline organic materials contained such elements suggested that wholly inorganic analogues could be made. The principles which enable high aspect ratio lamellae to be organised to form strong films and coatings has now been established (21).

Layer minerals, as their name implies, are composed of crystals which have a large surface area compared with their thickness. Crystallographic analysis (19) shows these structural elements result from the molecular organisation shown in figure 10. The unit crystal consists of an inner core of magnesium oxide/hydroxide which is connected to an outer silicate layer. This crystalline sandwich is 9 A thick with a surface area that is very large in relation to the thickness. An additional feature to the surface of the crystal is that in the polymeric Si-O layer, some tetravalent Si atoms are occasionally replaced by trivalent Al atoms, and more rarely by trivalent Fe-atoms, the effect of which is to make the surface negatively charged. This is balanced by an interlayer between lamellae having magnesium ions, with associated water molecules, so that overall the system is electrically neutral. This interlayer is 5 A thick and the

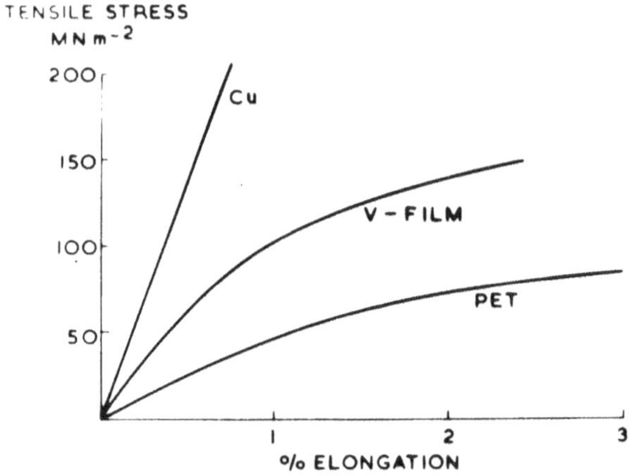

Figure 12

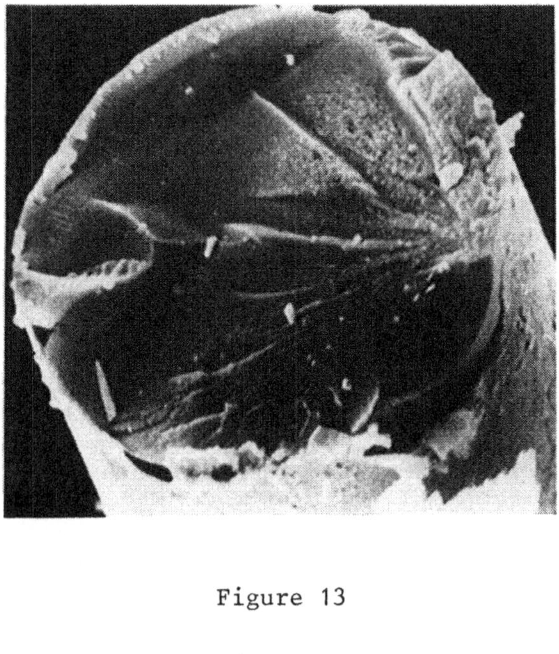

Figure 13

structure repeats itself thereafter every 14 A. The lamellae are stacked together like a pack of playing cards.

Several methods have been proposed to dissociate the macroscopic crystal into its fundamental lamellae (20). The preferred method employs the ion exchange properties of the magnesium gegen ion in the interlamellae regions. A two stage procedure consisting of exchanging magnesium ions for sodium in concentrated brine solution followed by replacement of the latter with an organic cation.

The effect is to produce a 30 fold expansion of the crystal to give a water based gel. This effect is shown in the photographs in figure 11 which show the dimensions of the initial crystal and the expanded gel produced. Within the gel total separation of the lamellae has not occurred but the process can be completed by shearing an aqueous dispersion of the gel in a colloid mill. Transmission electron micrographs (TEM) of the lamellae viewed normal to the surface show that they are several microns in diameter and are exceedingly thin, thin enough to appear transparent even where lamellae overlap.

It has been found that many of the properties of films prepared from these aqueous suspensions are determined by the distribution of particle thicknesses and particle diameters. These two parameters can be measured using a combination of techniques including micron sieving, dye adsorption and centrifugal sedimentation.

The aqueous suspension of lamellae is highly anisotropic and the phenomenon of streaming birefringence is observed on stirring or pumping through glass tubes. Moderate shear fields are sufficient to completely align the high aspect ratio lamellae. The forces exerted by surface tension on simply evaporating the water from the suspension produces total alignment of the particles and flat, highly coherent, films are obtained. These films vary in properties with the homogeneity of the suspension of the lamellae. There is also considerable variation in the optical properties of the films and clear, almost transparent, material can be obtained with very homogeneous suspensions. Some of these better films have properties which are superior to films obtained from organic polymers such as polyethylene and polyethylene terephthalate (PET). A typical stress-strain curve is shown in figure 12.

The surprising aspect of the stress strain curve is its close similarity to the organic films notwithstanding the fundamentally different molecule structure. There is, however, no recovery if the stress is released in the Hookian region whereas the PET film will recover some of the extension on releasing the stress.

The vermiculite film can also be formed around fibres or on metal surfaces. In many instances the adhesive is exceptionally good. A particularly useful application is the composite system obtained by encapsulating glass fibre as single filaments, or as woven fabric. In figure 13 is shown an SEM of glass fibres which have been coated with the aqueous suspension and dried. The unusual feature to the coating is that it totally encapsulates each fibre. Although the coating is 1 to 2 x 10^3 Å thick it consists of a least 100 lamellae arranged in a continuous layer. The latter protects the glass from fusion on heating above its melting point. It is an excellent substitute for asbestos without any of the toxic hazards.

Concluding Remarks

In this paper I have attempted to give an account of the background to the discovery of polyethylene in the early 1930's. In this I have been helped by discussion with Sir Michael Perrin. The success of the Winnington group lead to a growth in polymer science and publications from Northwich played an important part in developing the physical chemistry of polymer systems. Later they were joined by Charles Bunn, a crystallographer who defined the structure not only of polymers made by ICI but also the new polymers emanating from DuPont's Experimental Station. By the end of the 2nd World War polymer science was an established subject and polyethylene, together with Nylon, PET and others were part of the thermoplastic revolution in materials technology.

PHOTOGRAPHS

PLATE NO 1	F A Freeth as a young army officer.
PLATE NO 2	ICI's Winnington Laboratory erected in 1928. The statue on the far right is of Ludwig Mond FRS one of the founders of metal organic chemistry with his synthesis of the metal carbonyls.
PLATE NO 3	E W Fawcett.
PLATE NO 4	R O Gibson.
PLATE NO 5	M W Perrin.
PLATE NO 6	W R D Manning.
PLATE NO 7	High Pressure Apparatus typical of the time. Located in an open laboratory.
PLATE NO 8	Part of R O Gibsons note book showing the formation of a waxy solid on Wednesday 29 March 1933.
PLATE NO 9	Michael Perrins experiment on the 20 December 1935.
PLATE NO. 10	Typical reactor design used in the mid 1930's.
PLATE NO. 11	Page of the first sales ledger for Polyethylenes.

Plate 1

Plate 2

Plate 3

Plate 4

Plate 5

Plate 6

Plate 7

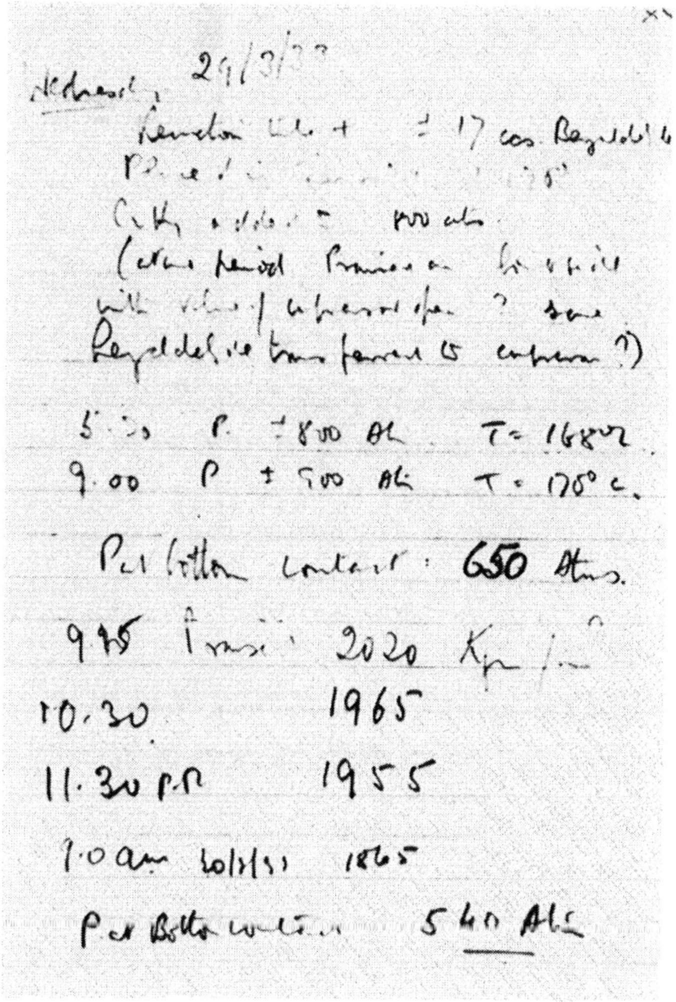

29/3/33

5.30 P = 800 Ats T = 168°C.
9.00 P = 900 Ats T = 170°C.

Put bottle contact : 650 Atms.

9.45 Pressure 2020 Kg/cm²
10.30 1965
11.30 p.m 1955
1.0 am 30/3/33 1865

Put Bottle contact 540 Ats

Plate 8

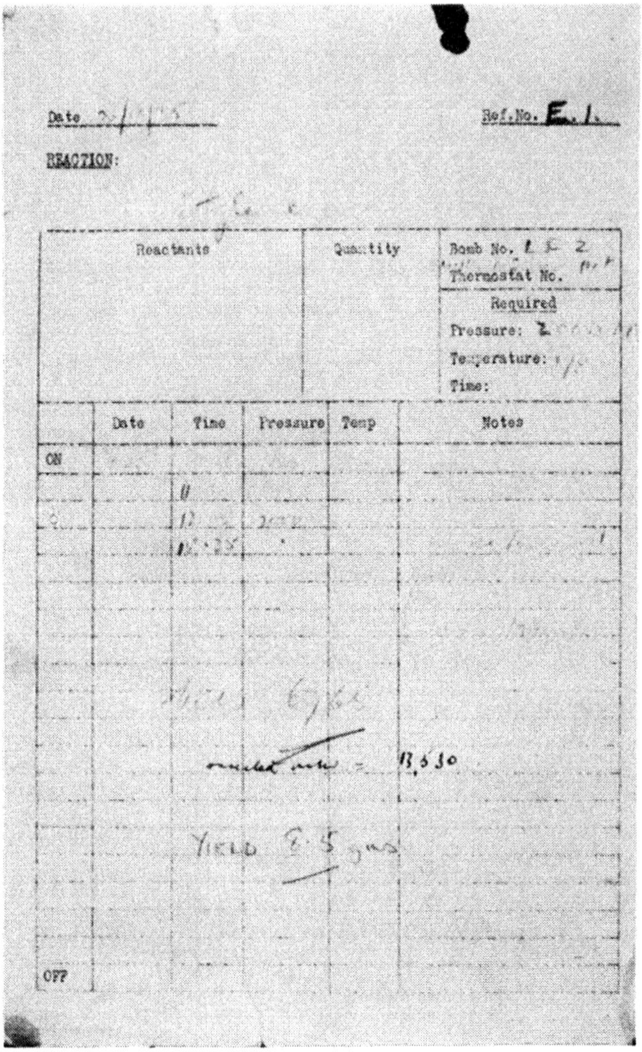

Date

Ref.No. E.1.

REACTION:

Reactants	Quantity	Bomb No.
		Thermostat No.
		Required
		Pressure:
		Temperature:
		Time:

	Date	Time	Pressure	Temp	Notes
ON					
OFF					

13,530

YIELD 8.5 gms

Plate 9

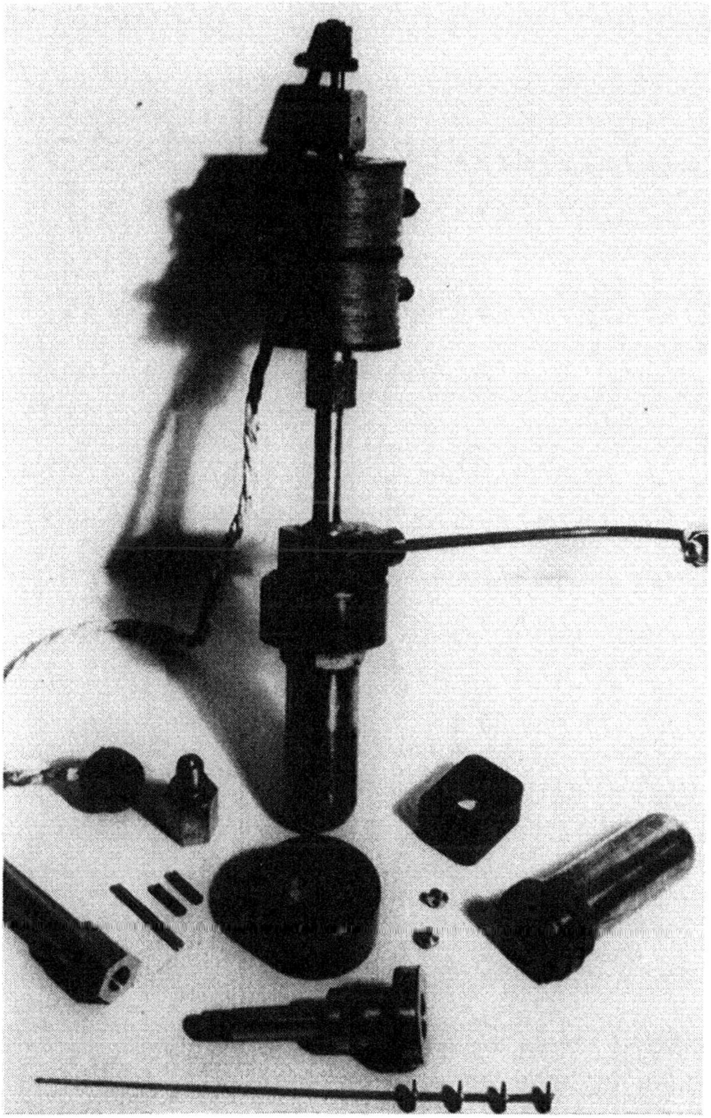

Plate 10

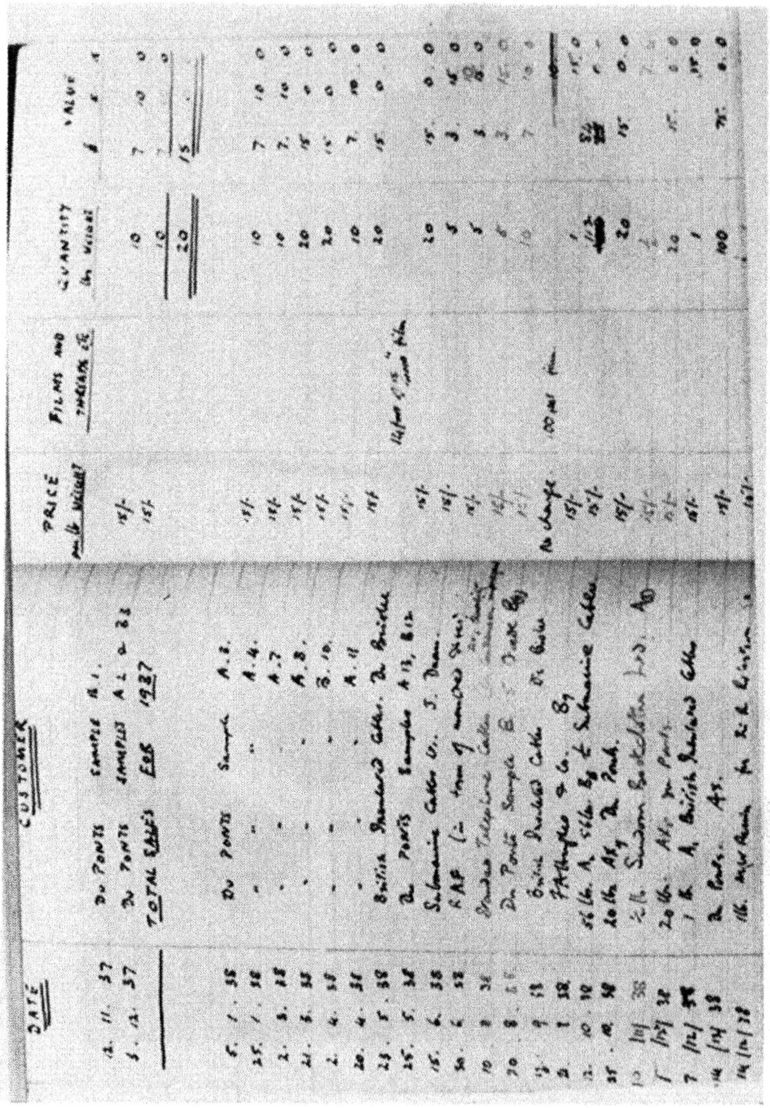
DATE
CUSTOMER
PRICE per lb weight
FILMS AND THREADS etc.
QUANTITY lbs weight
VALUE £ s d
12. 11. 37 Du Ponts Sample A.1.
5. 12. 37 Du Ponts Samples A 2 & B3
TOTAL SALES FOR 1937

Plate 11

Legends to Figures

1 The effect of pressure on k_p and k_t chain propogation and termination velocity constants in radical polymerisation of styrene at 30°C (Reference 7).

2 X-ray structure of Polyethylene. Ref (8).

3 Fringed Micelle Model. Ref. (10).

4 Single crystal of Polyethylene. Ref (11).

5 Isotropic polyethyene - a type of model which agrees with SANS studies. Refs. (12 and 13).

6 Drawn Polyproplene (8:1) showing the anisotropic distribution of the molecule in space. Ref. (14).

7 Annealed density of Polyethylene copolymers as a function of co-monomer content (a) propylene (b) hexane. Ref (15).

8 Two ways side chains can be distributed between crystalline and amorphous phase in polyethylene.

9 Comparison of various types of polyethylene copolymers.

10 Structure of the vermiculite mineral. Ref (19).

11 30 fold expansion of ion exchanged vermiculite (a) naturally, (b) after several hours in the same modium. Rof (21).

12 Comparison of stress strain curves for vermiculite film with copper foil and PET film. Ref (21).

13 SEM of fracture surface of glass fibre 15 μm diameter showing layered structure of coating. Ref (21).

References

1 Imperial Chemical Industries Ltd, A History by W J Reader, Oxford University Press, Vol II p.349, 1970.
2 E.W. Fawcett and R O Gibson, J. Chem. Soc. p.386, 1934.
3 M.W. Perrin, Discussion of the Faraday Society 34, 144, 1938.
4 Private discussion between D G H Ballard and M.W. Perrin, March 1985. A General Discussion on the Phenomena of Polymerisation and Condensation, Faraday transaction 32 1936.
5 J.C. Swallow Endeavour 3, 26, 1944.
6 M.G. Evans and M Polany, Trans Faraday Soc, 31, 875, 1935; Ibid, 32, 133, 1936.
7 K.E. Weale "Chemical Reactions at High Pressures", p.206, published by E&F Spon Limited, London, 1976. A.E. Nicholson and R.G.W. Norrish Disc, Faraday Soc, 22 97 and 104, 1956.
8 C.W. Bunn, Trans Faraday, Soc, 35, 482, 1939.
9 J. Hengstenberg, Z. Krist, 67 583, 1928.
10 P.H. Geil "Polymer Single Crystals", p.8, Interscience Publishers New York 1963.
11 P.H. Till, J. Poly. Sci. 24, 301, 1957; A Keller Phil, Mag, 2, 1171, 1957; E W Fischer, Z Naturforsch 12a, 753, 1957.
12 M. Stamm, E.W. Fischer, M Dettonmaier and P. Convent Faraday, Discussion of the Chemical Society, No 68, 163, 1979; D.G.H. Ballard, A.N. Burgess, T.L. Cowley, G.W. Longman and J. Schelten, Ibid 179, 1979.
13 D.Y. Yoon and P.J. Flory, Faraday Discussions of the Chemical Society No 68, 288, 1979.
14 D.G.H. Ballard, J Schelten, P Cheshire, E Janke, A Nevin, Polymer 23 1875 (1982).
15 D.G.H. Ballard and J. Pagett (1970) - unpublished.
16 E.W. Fischer and J. Schelten
17 B. Crist, W.W. Graessley and G.D. Wignall. Polymer 23, 1963 (1982).
18 G.D. Buckley and N.H. Ray, J. Chem. Soc., p.3701,(1952).
19 R.W. Grimshaw, The Chemistry and Physics of Clays, Fourth Edition Earnest Benn Limited, London 1971.
20 G.F. Walker and W.G. Garrett, Science 156, 385, (1967).
21 D.G.H. Ballard, G.R. Rideal, Journal of Mats. Sci. 18, 545, 1983.
22 U. Breitmeir and A.I. Bailey, Surface Science 89, 191, 1979.
23 A.I. Bailey and H. Daniels, J. Physical Chemistry 77, 501, 1973.

Appendix 1

Extracts from a report by M W Perrin written in ICI's Winnington Research Department of Alkali Division. Published 29 September 1936.

Figure 1 summarises the results of high pressure on the kinetics of "fast" and "slow" reactions.

The references display the chronological sequence of events leading to the final experiments on the polymerisation of ethylene in December 1935.

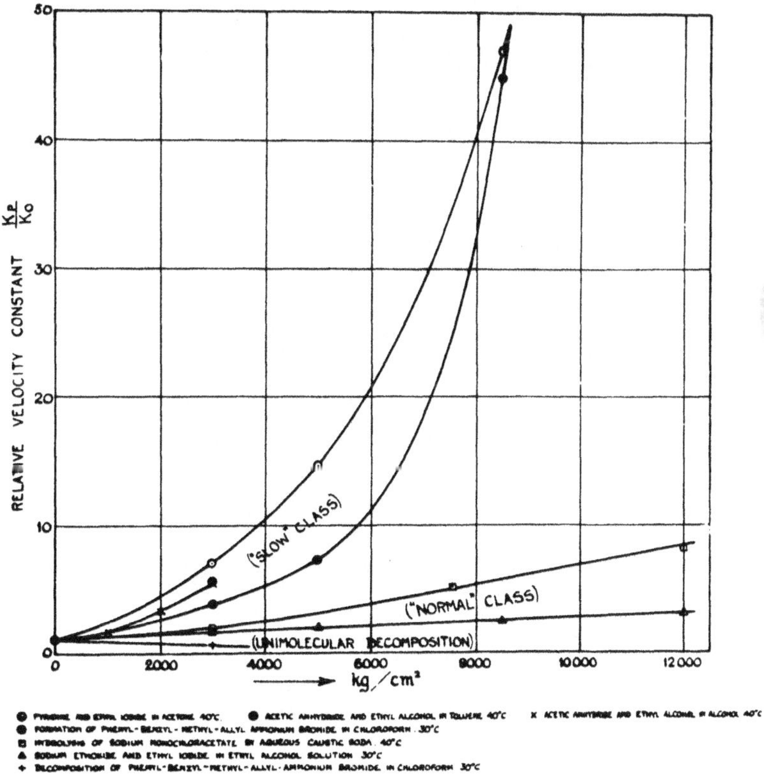

REFERENCES

(1) Chemistry at High Pressures. M.W. Perrin, J.C.Swallow. January, 1932. Central File No.R.621/2/1.

(2) High Pressure Research 1931-1935. J.C.Swallow, M.W.Perrin. May, 1935. Central File No. W.621/2/1.

(3) The Design and Construction of a 12,000 atm. Apparatus. W.R.D.Manning. June 1933. Central File No.R.621/5/3.

(4) "Engineering", 1933, 136, 32.

(5) The Technique used for Conducting Chemical Research at Pressures up to 3000 atms. W.R.D.Manning. March 1936. Central File No.R 621/5/5.

Apparatus and Methods used for the Study of Gas Reactions up to 3000 atms. R.O. Gibson, M.W.Perrin, W.R.D.Manning. December 1935. Central File No.R 621/2/14.

(6) The Effect of Pressure on the Binary System Monomethyl Aniline - Dimethyl Aniline. R.O. Gibson, J.C. Swallow. October, 1931. Central File No.R.736/1/1. J.C.S. 1934, 53, 18.

The Effect of Pressure on Binary and Ternary Systems. R.O.Gibson, December 1932. Central File No.R.666/2/1.

(7) The Effect of Pressures up to 3000 atms. on Chemical Reactions in the Liquid Phase. E.W.Fawcett, R.O.Gibson. August, 1932. Central File No. R.621/2/2.

The Effect of Pressure on Chemical Reactions. Second Interim Report. E.W.Fawcett, R.O.Gibson. May, 1933. Central File No R.621/2/4.

The Effect of Pressure on Chemical Reactions - The Formation of Cetyl Pyridinium Halides. E.W.Fawcett, R.O.Gibson. May, 1933. Central File No.R.621/2/3.

The Effect of Pressure on Chemical Reactions. II. Consideration of Results and Suggestions for further Study. E.W.Fawcett, R.O. Gibson. November, 1933. Central File No. R.621/2/7.

The Influence of Pressure on a Number of Organic Reactions in the Liquid Phase. E.W.Fawcett, R.O.Gibson. December, 1933. Central File No.R.621/2/8. J.C.S. 1934, 53, 386.

(8) The Synthesis of Formic Acid from Carbon Monoxide and Water at Pressures up to 3000 atms. M.W.Perrin, R.O.Gibson. September, 1935. Central File No.R.341/4/2.

(9) The Polymerisation of Unsaturated Hydrocarbons. E.W.Fawcett. February, 1936. Central File No.R.181/2/1.

The Polymerisation of Ethylene under High Pressure. M.W.Perrin, J.G.Paton, E.G.Williams. February, 1936. Central File No.R.181/2/2.

The Polymerisation of Ethylene at High Pressure. Preparation, Properties and Possible Uses of "Alketh". M.W.Perrin, E.G.Williams, J.G.Paton, C.W.Bunn. June 1936. Central File No. R.181/2/3.

(10) The Polymerisation of Linseed Oil under High Pressure.
E.W.Fawcett, R.O. Gibson, M.W.Perrin. June, 1935.
Central File No. R.123/35/1.

Preliminary Experiments with a Continuous High Pressure Plant for the Polymerisation of Linseed Oil.
W.R.D.Manning, M.W.Perrin. August, 1935.
Central File No. R.123/35/3.

Continuous High Pressure Polymerisation Plant. Results with Linseed Oil. M.W.Perrin, W.R.D.Manning.
February, 1936. Central File No.R.123/35/6.

(11) A. Michels. Proc. Roy. Soc. 1930, 127A, 258.

(12) For summary and references see: "The Quantum Mechanical Treatment of Energy of Activation of Chemical Reactions.
M.W.Perrin, September 1932. Central File No.R.276/5/2.

(13) The Effect of Pressure on Reactions in Solution.
R.O.Gibson, E.W. Fawcett, M.W. Perrin.
Proc. Roy. Soc. 1935, 150, 223.
E.G.Williams, M.W.Perrin, R.O.Gibson.
Proc. Roy. Soc. 1936, 154, 684.

(14) The Effect of Pressure up to 12,000 atm. on the Polymerisation of Linseed Oil.
M.W.Perrin, E.G. Williams, Jan.1936. Central File No. R.123/35/5.
M.W.Perrin, E.G.Williams, May 1936. Central File No. R.123/35/8.

(15) C.N. Hinshelwood. J.C.S. 1935, 1111.
Trans. Far. Soc. 1936, 32, 970.

(16) M.G. Evans, M. Polanyi. Trans. Far. Soc. 1935, 31, 875.
" " Trans. Far. Soc. 1936, 32, 1333.

(17) The Biological Effects of High Pressure. Summary of Work Published by J.Basset. M.W.Perrin.
March 1935. Research Dept. File No. Q.89/1/81.

See also Royal Society Discussion on "Methods of Measuring and Factors Determining the Speed of Chemical Reactions."
Proc. Roy.Soc. B. 1934, 116, 185.

Appendix 2

Sent to Dr. Slade 22nd Oct/36.
as a summary of what he was told during his visit (on 17th & 18th October)

17

Mr. J. C. Swallow

Research Department,
Winnington.

22nd October 1936.

A L K E T H.

PREPARATION.

A careful study has been made of the factors involved in the polymerisation of ethylene under high pressures. The reaction is exothermic and, if the temperature is allowed to increase above a certain value, an alternative "explosive" decomposition of ethylene to carbon and methane takes place. It has been found that a series of typical polymerisation catalysts have no marked effect on the reaction, but that the presence of very small amounts of oxygen causes a very rapid polymerisation of a certain proportion of the gas. This rapid reaction is followed by a slow steady polymerisation which probably takes place in the absence of oxygen.

Experiments have been carried out in which ethylene, with a suitable oxygen content, has been passed rapidly through a heated reaction space to give a 10-15% conversion to polymer and in which the gas has been polymerised by a static method to give yields of about 70%.

PILOT PLANT.

The possibility of making a mechanical gas compressor to work with ethylene at pressures above 1000 atmospheres was discussed with Michels who produced an original design. This has been submitted to Major Bramwell and to Brown (Billingham) for comments. It was decided in April that this compressor should be constructed in Holland under Michels' direction. It will have a maximum capacity of 5 M^3/hour (= 50 tons C_2H_4/year) with a filling pressure of 200 atms.

Reaction vessels of 3/4 L. and 9 L. capacity have been designed and were ordered at the end of July.

It is hoped that the complete plant will be ready during December, and it should have an initial production rate of about 1 lb./day.

FULL-SCALE PLANT.

Preliminary schemes have been considered for the production of 100 tons Alketh per annum by a continuous and a batch process, and a tentative cost estimate prepared. On this scale the conversion cost may be expected to be between 6d and 1/6d per lb. while ethylene (from ethyl alcohol) will cost from 3d to 5d./lb.

PATENTS.

A series of provisional patents have been filed (the first dated 4/2/36) covering the production of the polymer with controlled rate of heat removal and controlled oxygen content, and the variation of the molecular weight of the polymer with pressure and oxygen content. In addition the production and uses of the polymer in the form of threads, staple fibre, films, tape and moulded articles have been covered.

USES.

It is felt that, of a large number of possible uses, attention should first be concentrated on those connected with the electrical industry, where the outstanding insulating properties of Alketh combined with its flexibility in the form of tape and films, and its chemical inertness,would appear to be most promising.

(Signed).... M.W.Perrin.

J.P. MACHON

RESEARCH WORK IN HIGH PRESSURE POLYETHYLENE SYNTHESIS AT C.d.F. CHIMIE

SYNOPSIS

A general survey on the research work conducted by C.d.F. CHIMIE in the high pressure polyethylene synthesis is given for these last thirty years. Studies carried out on LdPE, LLdPE and radical copolymers are briefly described. The future developments concerning the research in the high pressure polyethylene synthesis are also outlined.

INTRODUCTION

The purpose of this paper is to give a survey on the research progress at C.d.F. CHIMIE in the high pressure synthesis of ethylene polymers.

Thus, the large steps which characterized C.d.F. research efforts in the field of LdPE, LLdPE and other copolymers will be presented while the principal topics of research studies will be described. The aspects for evaluation and improvement of polymer are not presented in this paper.

1. HISTORY OF C.d.F. RESEARCH WORK IN THE FIELD OF POLYOLEFINS

The first industrial polyethylene plant was started in 1954 at MAZINGARBE in the north in France on the basis of ICI licence.

. Starting of the research politic : organization of means (1955-1965)

The first efforts have consisted in giving the means in transformation machines and measurement apparatus to evaluate commercial grades and experimental products.

An application laboratory well equipped with different machines was built in 1958 at MAZINGARBE to examine commercial grades by injection-molding, extrusion coating, tubes techniques. At the same period, a chemical and physical laboratory an

R. B. Seymour and T. Cheng (eds.), History of Polyolefins, 55–69.

discontinuous pilot plants were also built at MAZINGARBE. Objectives were to study polymerization, structural and chemical properties of polyolefins like polyethylene,polypropylene and polybutene.

All the persons concerned by polyolefins R and D amounted to more than one hundred and fifty. During this period, beside the evaluation of commercial grades and technical assistance given to the customers, the most important parts of researches were :

* Ziegler low pressure olefin polymerization
* Radical high pressure polymerization
* Caracterisation of macromolecular structures by solution techniques (GPC, viscosimetry, light diffusion)
* Evaluation of the polymer by spectroscopic, microscopic techniques and by molten-state rheology
* Transformation techniques and stabilization.

This period is characterized by studies with a basic character to build the scientific and technical knowledge on kinetics and polymerization mechanics.

Meanwhile people working on typical research problems and on process problems organized themselves in a Research Department and a Process Department. The Process Department was more specially oriented towards the specific and technical aspects of the high pressure plants.

. **Maturity phase : research is oriented towards the improvement of LdPE high pressure process (1965-1976)**

A continuous pilot plant was equipped in 1965 to allow experiments in the field of the high pressure polymerization of ethylene. Progressively, between 1967 and 1976 other pilot plants with various sizes (reactor volume : 1 l - 6 l - 50 l) were built to do the industrial transposition of experimental results obtained for kinetics, initiators and transfer agents or comonomers.

Parrallely, the growth of industrial plants (from 2,500 T/year to 60,000 T/year) drives the Process Department to investigate the following technological aspects :

* reactor design, separators and side equipments, high pressure and medium pressure security devices
* design and supply of h.p regulation valves, H.P flowmeters
* study of flowing for ethylene and ethylene-polymer mixtures.

During the same period, investigations on the polybutene led to perfect an original mass process which wasn't industrialized. Investigation on polypropylene were also stopped because C.d.F. CHIMIE didn't want to enter these two product markets. These decisions led to a reinforcement of research efforts to perfect an ionic high pressure process for the manufacture of hd polyethylene.

. Development phase (1976 - 1985)

This last period expressed a will of diversifying the manufactured products in preserving C.d.F. technology and know-how in high pressure polymerization. The basic ideas for the manufacture of LdPE was that of the conception of very large industrial plants (150,000 and 180,000 T/year) so that the operative costs would be decreased to the maximum.

Progress accomplished in the technology field permitted this fact. At last, the ionic high pressure process started to work efficiently.

In this period, C.d.F. CHIMIE put different plants into service :

- May 1978 : industrial launching of a retrofitting line (30,000 T/year) to manufacture HdPE at LILLEBONNE
- November 1978 : starting up of a new 180,000 T/year for LdPE at DUNKIRK
- March 1980 : industrial launching of LLdPE on two industrial lines at LILLEBONNE
- November 1983 : starting up of a new 100,000 T/year line for LLdPE at DUNKIRK
- March 1984 : industrial launching of ethylene/acrylic/ maleic anhydride terpolymers.

During that period, research costs were also compressed ; savings were searched and operative costs of big pilot plants becoming too expansive they were stopped. Actually, research teams which are concerned by polyethylenes and copolymers amount to about ninety scientists.

2. RESEARCH WORK CONDUCTED IN THE HIGH PRESSURE POLYOLEFIN SYNTHESIS

From a general manner studies carried out on mixing effects for the quality of the polymer in the sixties haven't given any discounted results

Fig. 1

POLYMERIZATION WITH NITROGEN : TREF DISTRIBUTION

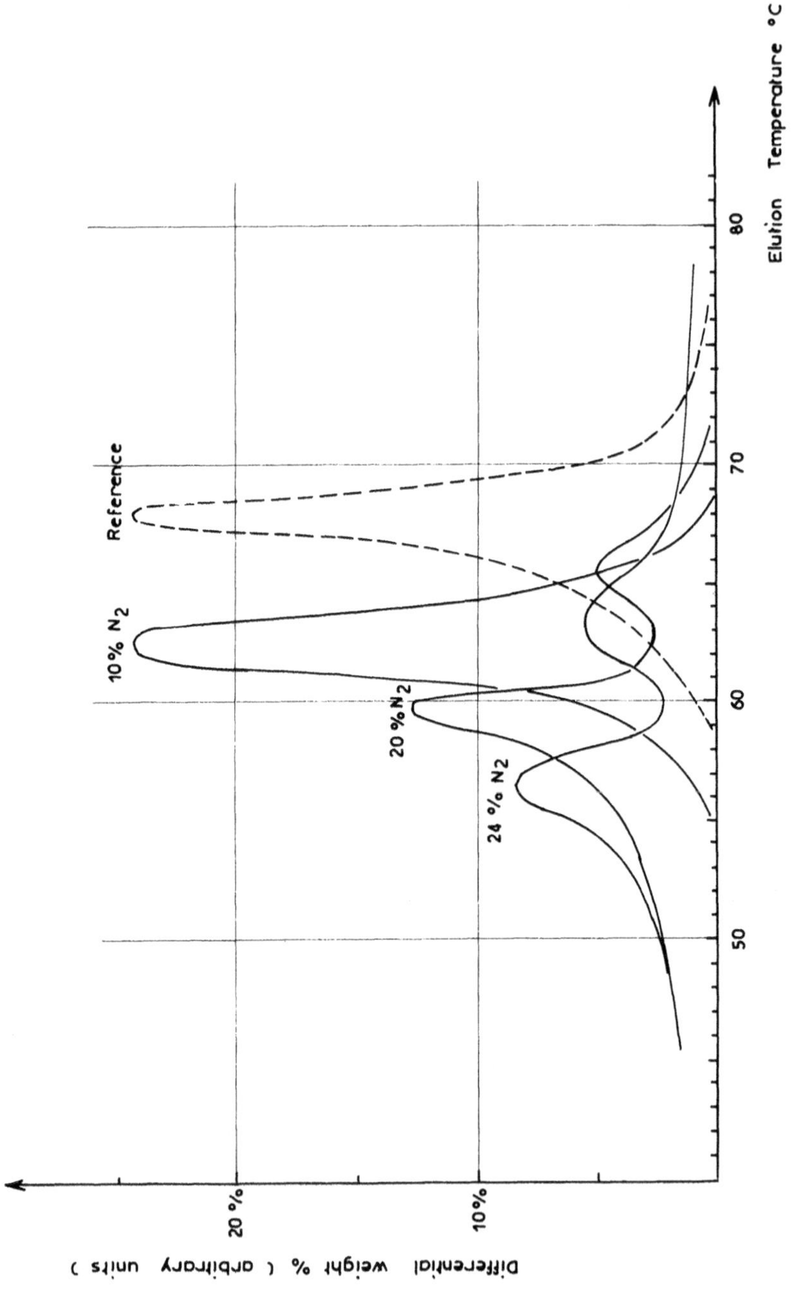

Fig. 2 POLYMERIZATION WITH NITROGEN : MOLECULAR WEIGHT DISTRIBUTION

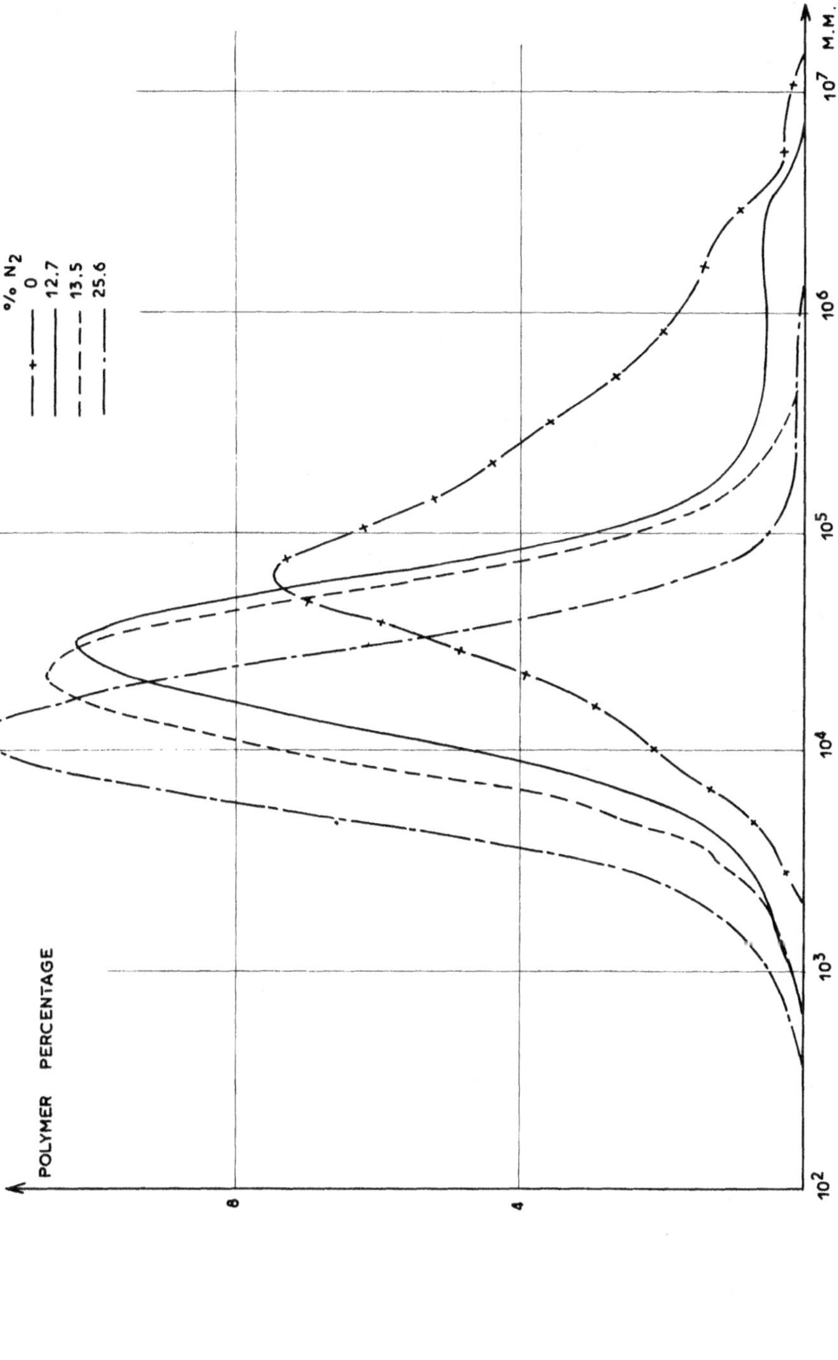

2.1. RADICAL POLYMERIZATION : LdPE AND COPOLYMER SYNTHESIS

The importance of research work done in the high pressure radical polymerization of ethylene by some scientific teams must be outlined. Thus : P. EHRLICH and G.A. MORTIMER (MONSANTO), Th. VAN der MOLEN (D.S.M.), R. STEINER (HOECHST), G. LUFT (Institut für CHEMISCHE TECHNOLOGIE) have conducted essential experiments on the following aspects : kinetics and microstructure of polymer, initiator efficiency, transfer - agents and comonomers.

For its own part, on top of the classical studies quoted above, C.d.F. CHIMIE has conducted original investigations in studying possibilities to adjust the structure of the polymer from the state of reaction mixture. As others, we have searched to predict monomer cunsumption and structural characteristics of macromolecules from the elementary steps of the kinetics. A new approach for modeling has been attempted.

2.1.1. Effects of the phase separation during polymerization

We have already communicated a part of these studies (1). Different ways for producing phase separation have been explored and only some aspects will be remindered in this paper.

* Dilution with nitrogen

Phase separation occurs as soon as small amounts of nitrogen are added to the reaction mixtures (2) : this phenomenon is confirmed by stirring power observations, by the bimodal short chain branching distribution and by the narrowing of the molecular weight distribution, when nitrogen concentration increasing (see figures 1 and 2).At that time we concluded : althrough several arguments for phase separation exist, the radical polymerization of ethylene performed with nitrogen does not induce, as expected, formation of macromolecules with increasing long chain branching when nitrogen concentration increases. The opposite effect is observed. It is probable that a decreasing ethylene concentration involves an increasing free radical concentration in order to maintain a constant polymerization rate. The rate of intramolecular transfer reactions leading to short chain branching increases ; after transfer reactions giving secondary and tertiary carbon radicals, the β-scission of these radicals should lead to more vinylidene and vinyl unsaturations and therefore to less long chain branching.

From this time, owing to the GPC viscosimetry coupling

Fig. 3 POLYMERIZATION WITH WATER : TREF DISTRIBUTION

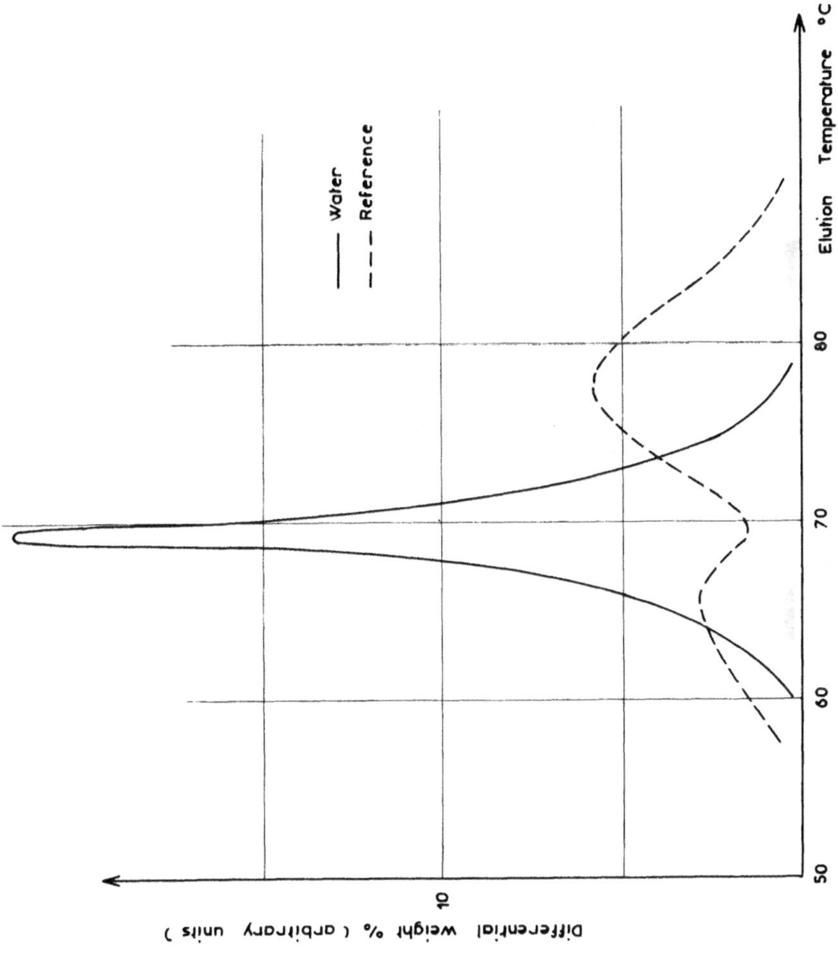

we have seen that very important effects are observed on long branching frequency repartition. Recent studies have showed that polyethylene grades produced with nitrogen dilution present excellent properties, specially high optical properties.

* Dilution with water

By analogy with water-ethane system, we have supposed that, in the main part of the concentration range, a water rich phase and an ethylene rich phase (organic phase) already exist before polymerization. Polymerization takes place essentially in the organic phase because the initiators are soluble in it. The organic phase during polymerization in presence of dispersed water can itself undergo phase separation under the effect of temperature, pressure polyethylene concentration and dissolved water. Polymerization in the presence of water, if compared to the reaction in the same conditions without water, yields higher ethylene conversion (3) ; polyethylenes produced are very different because in the second case the temperature rising elution fractionation shows a bimodal distribution which is characteristic of a two phase polymerization. For the sample polymerized in presence of water this characteristic vanishes and the SCB distribution gets sharper (figure 3). Disappearance of two phases can be explained by increase of PE concentrations which produces a shift from the heterogeneous to homogeneous region of the phase diagram of PE-ethylene at a given temperature and pressure.

All these effects are consequent to conversion increase rather than specifical effects of water on physical-chemical behavior of the organic phase.

. Effects of pressure, temperature and conversion

These parameters have been extensively studied in industrial reactors and their effects on molecular weight or short chain branching distributions are well known.

We are now convinced that phase separation is a very useful polymerization parameter able to produce considerable structure modifications especially on molecular weight and short chain branching distribution which determine applications properties as optical properties and stretchability. It is well-known that an important part of autoclave grades are now produced in demixtion and performances of these grades are comparable to tubular grades.

2.1.2. Copolymers : improved polyethylene grades

Like many companies we have searched comonomers able to bring stiffness and adhesion properties. After a lot of experiments with various comonomers (1970-1980 period) we have finally chosen an ethylene-acrylic ester-maleic anhydride ternary system. Ethylene-acrylic ester binary copolymers are well known (4) and have been commercialized for a long time by a few companies ; they present some advantages (better thermal stability, better low temperature mechanical properties) against vinylacetate-ethylene copolymers (EVA) which have a low thermal stability. This instability limits the conversion in polymerization or the output in extrusion.

The acrylic esters don't present this drawback and their copolymerization reactivity is bigger. We have finally chosen ethyle and butyle acrylates which, by the volume effect of the lateral group, have similar effects to octene in an olefinic copolymerization for the mechanical properties (figure 4).

The use of the maleic anhydride as termonomer enhances the adhesion properties and offers also the possibilities of posterior chemical treatments (5).

C.d.F. CHIMIE commercializes now these terpolymers (6) on a large scale of composition ; terpolymers with acrylic ester and maleic anhydride molar contents respectively below 10 and 1 % are manufactured with particular economical conditions between 1000 and 2000 bars and 180-260°C.The use of an autoclave reactor permits to do a good mixing of the comonomers which is indispensable to obtain homogeneous copolymers when comonomer reactivity is ten times as important as ethylene.To perfect these terpolymers an important part of C.d.F. research work has consisted in finding means to avoid too high consumption of initiators and decompositions linked to the presence of maleic anhydride.

2.1.3. Modeling ethylene radical polymerization

The term polymer quality means the set of numerous structural characteristics of macromolecules (molecular weight, molecular distribution, short chain and long chain branching ....). These structural characteristics control the engineering behavior of the polymer and its capacity for particular industrial applications. Thus, it is of paramount importance to be able to predict to what extent they depend upon the design and operation of the polymerization reactors.

During a long period (1970-1979) different approaches have been done without success : the two classical methods for linking the quality of the polymer to the operating conditions of the reactor, the "black box" approach relying on empirical correlations and the "detailed analytic" approach haven't permitted to link the structural characteristics of the polymer to

Fig. 4 EFFECT OF ACRYLATE TYPE ON DENSITY OF TERPOLYMERS

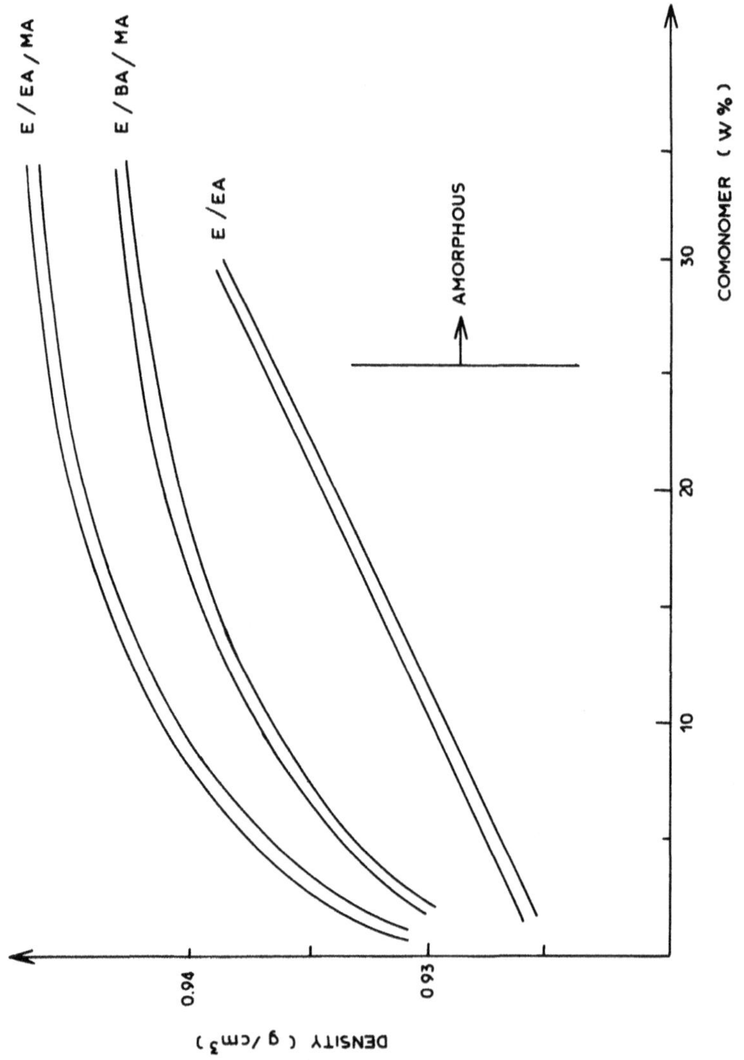

the operating conditions. More recently a new approach with a tendancy model (7) and new assumptions on the dissolution of the initiator in the monomer have permitted to explain initiator and monomer consumptions, reaction zone polymer production, structural characteristics of the polymer in homogeneous or demixed phase. The novelty of the model comes from the fact that concentrations of moments of the molecular weight distribution and of chemical characters are considered instead of individual molecular entities.

This tendancy model permits to simulate an industrial reactor in permanent steady pace and is useful for the transposition of Pilot experiments to industrial runs.

2.2. IONIC POLYMERIZATION

In the field of the ionic high-pressure polymerization, C.d.F. CHIMIE has started exploratory experiments from 1965. More important abilities were appropriated since 1970 and the ventured efforts emerged in march 1979 on the starting of a high pressure industrial plant (30.000 T/year) reconverted for the production of high density polyethylene.

Since that time, the research abilities have been concentrated towards the LLdPE and the manufacture of hd polyethylene has been substituted by the LLdPE since the beginning of 1981.

The transposition of the radical process to ionic process has required an important research work under various aspects. The most important are :

* high temperature Ziegler catalysis, kinetics and operating parameters
* studies on copolymerization with different comonomers and evaluation of the copolymers
* LLdPE grade improvements by creating long chain branching
* basic studies on phase equilibrias.

Only the first aspects are presented in this paper.

. High-temperature Ziegler catalysis

Because the use of high temperature Ziegler catalysts for which transfer and termination reactions have considerable importance, C.d.F. CHIMIE has devoted extensive efforts to investigate suitable catalytic systems.Our research has been marked by the fact that the main objective underwent a change during 1980 so that two periods can be distinguished :

* an initial period (1970-1980) during which efforts were concentrated on developing catalysts suitable for the manufacture of high and medium density linear polyethylene (0.94 to 0.965 g/cm^3) (8) ;

* a second period, from 1980, during which we have been looking for systems more suitable for copolymerization and production of low density polyethylene (9).

It may be remindered here that the catalyst is the main factor that can be used to affect :

* copolymerization (individual comonomer reactivities)
* inter and intra heterogeneities
* usual effects connected with various transfer and termination reactions (molecular weight distribution).

All these catalyst studies have led to perfect large family of binary titanium-vanadium systems which can be used with stabilized alkylaluminiums as pentalkylsiloxyalanes.There are various possibilities to optimize the composition of solid solutions of such binary catalyst systems (for example by altering the grinding temperature or by carrying out annealing and hardening operations) ; C.d.F. CHIMIE is pursuying investigations in this field.

Parallely to these investigations on catalyst, the study of the desactivation has required some efforts ; the catalyst is not completely inactive at the outlet of the reactor and it must be destroyed to avoid the plugging of recycles.

The study of operating parameters has showed that the influence of the pressure was not so important as temperature monomer and hydrogen concentration (10) (figure 5).

. <u>Copolymerization with different comonomers</u>

The use of propylene, 1.butene, 1.hexene allows to produce very various polyethylenes. The high pressure process presents undoubtedly appreciate advantages from this point of view :

* Copolymerization with 1.butene allows to encompass a large field of grades between the high density and 0.915.

* Terpolymerization with ethylene/propylene and butene allows to access to very low density polyethylene as low as 0.900 and presenting high toughness and flexibility. These two properties with their aptitude to reticulate make them very attractive for squeeze tubes and bottles, drum liners and cables.

* C.d.F. CHIMIE has also investigated the possibilities to produce ethylene/propylene elastomers with the high pressure process ; howewer the disappearance of the cristallinity requires the incorporation of about 45 % by weight of propylene in the copolymer.

Fig. 5 IONIC POLYMERIZATION

CATALYTIC EFFICIENCY VERSUS PRESSURE AND TEMPERATURE

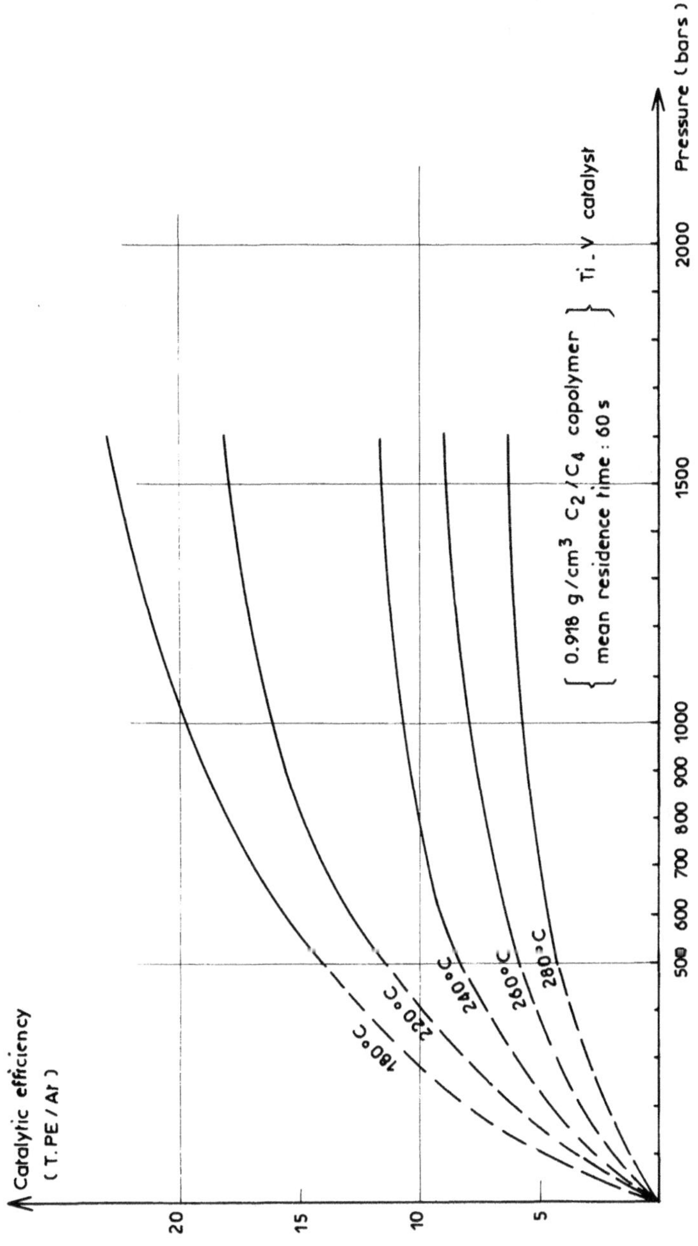

This content in the polymer needs with present catalysts, a concentration of about 70-80 % by weight of propylene in the feeding gaz. For the moment, the main difficulty proceeds from the polymer melt index which cannot be easily decreased and the use of special catalysts becomes required.

The use of heavy comonomers like 1.hexene brings real advantages when high stiffness and good mechanical properties are simultaneously required. The mass process permits to use liquid comonomers : on the reaction plan, the high pressure copolymerization of ethylene with hexene don't involve special problems but the reactivity of heavy olefin is diminushed noticeably. This fact requires to recycle large quantity of heavy olefins (comonomer conversion rate : about 4 %). The catalytic efficiency is little affected and ethylene/hexene copolymers can be produced with same catalysts that for butene copolymerization.

From a process point of view, it is necessary to optimize the separation pressure because the reciprocical solubilities of polymer and gaz are very different and change the respective recycling flows.

. LLdPE grade improvements :

It is now well-known that the presence of small amount of long chain branching exchanges the processability of LLdPE film grades and other properties like tear strength. C.d.F. has much studied the different possibilites to create long chain branching directly in line.

CONCLUSION

The most important part of the research work in high pressure polyethylene synthesis has had industrial effects in leading to :

* phase demixtion as a control means for LLdPE structure
* elaborating a process for the manufacture of ethylene/acrylic ester/maleic anhydride terpolymers
* elaborating a new high pressure process for the manufacture of linear polyethylene.

As well as the size increase of high pressure plants was a favorable element to reduce operating costs of polyethylene, it can be supposed that a new big improvement could arise from the possibility to automatize the plants completely and to optimize the quality/cost ratio.

At last, developments in the field of LLdPE are stimulating

for conventional polyethylene and it is very likely that new improved grades will be elaborated for LdPE ; new very low density linear polyethylenes (0.880 to 0.910 g/cm³) will be soon produced and will enhance the field of the LLdPE.

Anyway all the studies conducted in the field of the ethylene high pressure polymerization have brought a large know-how for CdF : the taking out of many patents and numerous licensees around the world are a good proof of it.

C.d.F. CHIMIE S.A. - Centre de Recherches
62670 - MAZINGARBE - FRANCE

BIBLIOGRAPHIE

(1) D. CONSTANTIN - M. HERT - J.P. MACHON
Meeting of the European High Pressure Research Group
September 1,2,3 (1980) - AMSTERDAM

(2) D. CONSTANTIN - J.P. MACHON
European Polym. J. - **14**, 703 (1978)

(3) D. CONSTANTIN - M. HERT - J.P. MACHON
European Polym. J. - **17**, 115 (1981)

(4) USP n° 2,953,551 - UNION CARBIDE (1960)

(5) J.C. DECROIX - J.M. BOUVIER - R. ROUSSEL - A. NICCO
C.M. BRUNEAU
J. Poly. Sc. Symp. n° **52**, 299 (1975)

(6) J.C. DECROIX - D. EMMANUELLI - J.P. MACHON
EUROPLASTIQUES (1982) - PARIS

(7) J. VILLERMAUX - L. BLAVIER - M. PONS
Polymer Reaction Engineering - HANSER PUBLISHERS
(1984)

(8) J.P. MACHON - EUROPLASTIQUES (1978) - PARIS

(9) J.P. MACHON - GOLDEN JUBILEE CONFERENCE -
PE 1933-1983 LONDRES (1983)

(10) J.P. MACHON
European Polym. J. - **12**, 805 (1976)

Malcolm Dole

THE HISTORY OF THE CROSSLINKING OF POLYOLEFINS

1. INTRODUCTION

The practical importance of the crosslinking of polyolefins was demonstrated by Charles Goodyear and others in the years 1838-1841 when it was discovered that natural rubber when heated with sulfur became much stronger and harder; in other words it was converted to vulcanized rubber. Now about a century and a half later sulfur is still the major vulcanizing agent.[1] Another crosslinking of polyolefins discovery of major industrial importance was that of the author[2] in 1948 who studied the effect of high energy rays on polyethylene (PE) and found that after an irradiation in the heavy water pile of the Argonne National Laboratory the stress strain curves of PE were greatly altered--the cold drawing ability of the PE had been completely removed.[2] This was interpreted in terms of the formation of C-C chemical bonds between the long PE chains, in other words in terms of crosslinks. This chemical bond formation occurs at room temperature without alteration of the shape of the plastic sample being irradiated. Many millions of dollars worth of heat shrinkable tubing and films based in part on this technique are now being sold each year in the United States and around the world. In addition to the practical importance of crosslinking reactions there is much information of theoretical interest to be gained by radiation studies as discussed below.

Because polyethylene has few ultraviolet light absorbing groups in it one would expect little if any practical interest in using UV light to crosslink PE, but in 1956-59 Oster and coworkers[3,4] using UV light in the 200-300

R. B. Seymour and T. Cheng (eds.), History of Polyolefins, 71–86.

mμ wave length range found that the rate of reaction was greatly increased by the incorporation of certain sensitizers into the plastic. Details of this work are given below. The practical importance of UV crosslinking may be in the surface grafting of monomers onto polyolefins such as the grafting of styrene onto PE.[4]

In addition to vulcanizing rubber with sulfur other chemical reagents can be used to crosslink polyolefins. For example, Van Der Hoff[5] in 1956 and later years showed that polyisoprenes could be vulcanized using dicumyl peroxide. Also Kovacic and Hein[6] in 1959 found that dimaleimides could crosslink natural rubber at high temperatures, but only to a slight extent. Addition of a catalytic amount of a peroxide to this system greatly increased the degree of crosslinking. The crosslinking of polyethylene by means of a mixture of a dimaleimide and dicumyl peroxide is mentioned below.

With this introduction we now turn to a more complete discussion of the items listed above.

2. CHEMICAL CROSSLINKING

As is well known, the chemical crosslinking of rubber by heating natural rubber with sulfur is of enormous importance, especially in the automobile tire industry. When the earliest settlers came to the Western Hemisphere in the late 15th century, they found that the Indians in Central and South America had learned to extract latex from rubber-yielding trees and to use it for waterproofing garments.[7] But as we all know natural rubber cannot retain its elasticity as the temperature is raised. Nevertheless, rubber was used to incorporate it into garters, etc. Thomas Hancock had a factory in London in 1820 and devised a masticator for dealing with scrap rubber. In 1838 Nathaniel Hayward found that on exposing a mixture of sulfur and rubber to sunlight the surface became nontacky.[8] He patented this process; the patent was purchased by Goodyear who then mixed rubber and sulfur and heated the mixture to produce vulcanization. He patented this process in 1841 and shortly after Hancock patented a similar process.[8] In 1932 Ludersdorff[9] in Berlin made small balloons formed from a hot solution of rubber in turpentine to which sulfur had been added. Stickiness in the finished product was avoided. There are a number of accelerators that can be used in the vulcanizing of rubber. For example in a book published in 1961 Sorenson and Campbell[10] describe the use of sulfur together with an accelerator, such as mercaptobenzthiazole, in the compounding of rubber. Carbon black is another useful filler.

The first curing of rubber with a peroxide was that of Ostromislensky in Russia about 1914 who used benzoyl peroxide.[11] Later in 1955 Amberg and Willis[12] found that dicumyl peroxide was a better peroxide to use. Patterson and Koenig[13] very recently studied the structure of natural rubber vulcanized with dicumyl peroxide and found that the networks contained only carboncarbon crosslinks, namely

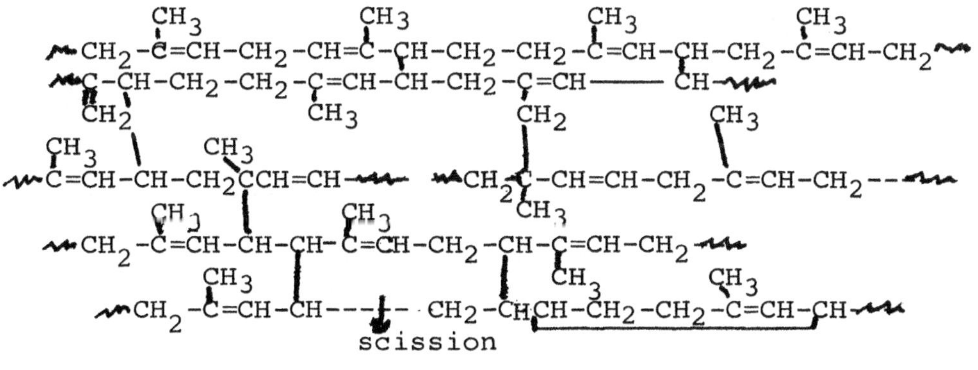

In Van der Hoff's 1956-63 studies of the mechanism by which polyolefins can be crosslinked by peroxides he suggests that in the case of dicumyl

peroxide the initial step in the reaction is the splitting of the peroxide into two free radicals, namely

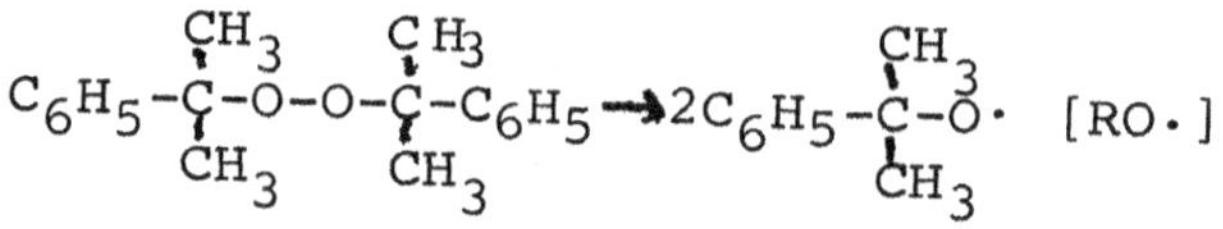
(1)

followed in part by the reaction

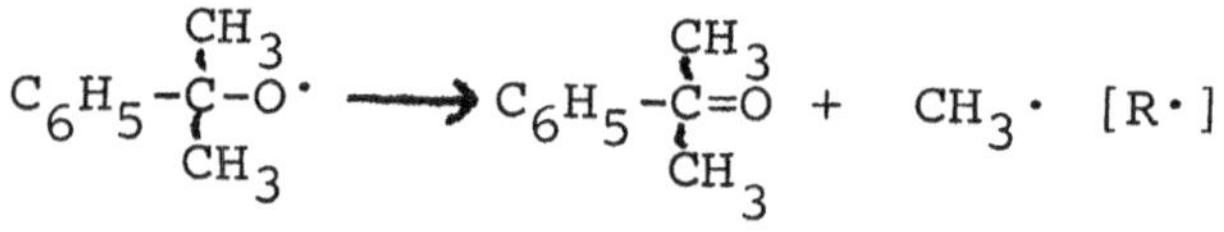
(2)

The free radicals RO• and R• can then abstract a hydrogen atom from the polyolefin to yield a polyolefin free radical

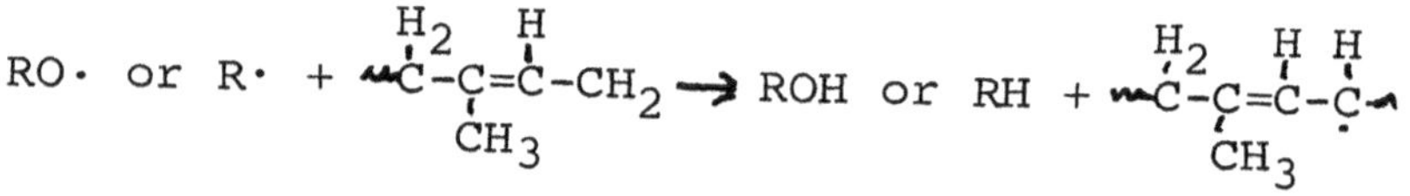
(3)

$$P\bullet + P\bullet \rightarrow P - P \qquad (4)$$

This mechanism results in one crosslink for each molecule of dicumyl peroxide decomposed. The crosslinks produced in natural rubber were proportional to the dicumyl peroxide concentration. In the case of curing cis-polybutadiene (cis-PB) the number of crosslinks increased up to a factor of 50 over that of rubber due to the chain reaction, eqs. 5 and 6

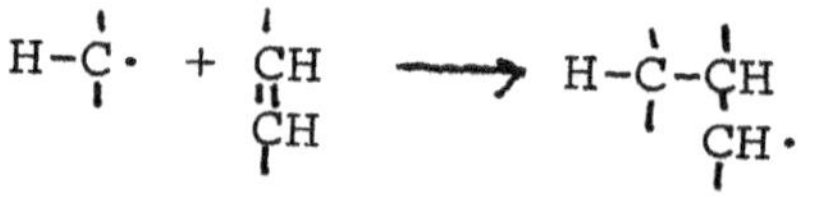
(5)

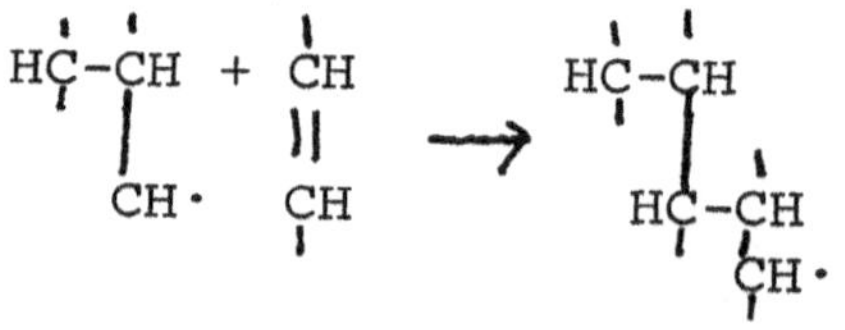
etc. (6)

It is interesting to note that in a recent study of the γ-ray initiated crosslinking of PB at -95 carried out by Basheer and Dole[14], higher yields than expected of crosslinks were obtained. These authors also explained this result in terms of a chain reaction similar to eqs. 5 and 6.

In addition to the above and at about the same time, (in 1959) as mentioned above Kovacic and Hein[6] reported on the "Crosslinking of Unsaturated Polymers with Dimaleimides." The suggested reactions were initiated by free radicals as in reactions 1 and 3 above, namely

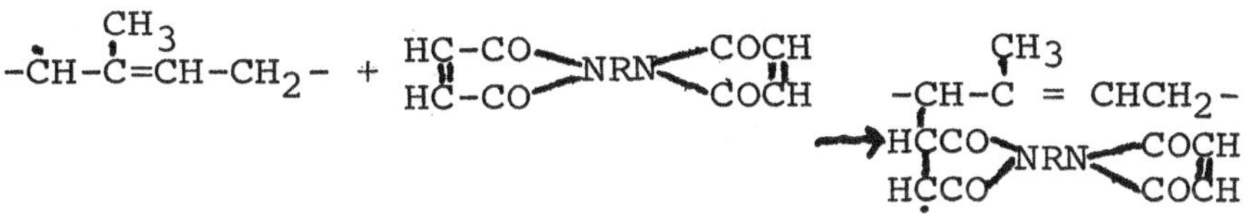

(7)

The latter free radical reacts further with the polyolefin to give

(8)

and then the process can repeat itself setting up a chain reaction. When the concentration of dicumyl perioxide or benzoyl peroxide was decreased in amount by 80% no insolubilization occurred.

Kovacic and Hein also found that they could produce gelation in low density PE when heated to 177° in o-dichlorobenzene with 4% of the dimaleimide and 0.4% of dicumyl peroxide present. In 1963 Precopio and Gilbert of the General Electric Company obtained a patent[15] for the crosslinking of polyethylene with dicumyl peroxide. Another chemical method of crosslinking polyethylene is that using polyethylene having alkoxysilyl groups grafted to the PE molecules.[16] The crosslinking reaction is

$$-\overset{|}{\underset{|}{Si}}-OR \;+\; RO-\overset{|}{\underset{|}{Si}}- \;\xrightarrow[\text{Catalyst}]{H_2O}\; -\overset{|}{\underset{|}{Si}}-O-\overset{|}{\underset{|}{Si}}- \;+\; 2ROH \qquad 9)$$

In 1960 Carlson[17] wrote a review in the "Rubber World" on "Compounding Cross-Linked Polyethylene" in which he pointed out that low-density PE resins have the lowest crystallization temperature and are, therefore, most readily adapted to processing in rubber equipment. Among the peroxides which could be used it is necessary to choose one which decomposes in the temperature range and cure times normally used in the compounding of rubber. Carlson recommended Varox (a trademark of the R. T. Vanderbilt Company) which contains 50% of di-t-butylperoxy dimethyl hexane in an inert carrier. Usually 4-6 parts of Varox are used per 100 parts PE. The greater the crosslinking the less the percent of compression as measured under constant load for 30 minutes at 120°. Carlson recommends this compression test as a convenient method for relating properties of the polyethylene to the degree of crosslinking. For example, after curing of the PE for 120 minutes at 160° the percent compression was reduced from 80% to 8.1%. Some antioxidants can be used to protect the PE, but many retard the cure rate. One which has a minimum adverse effect on the state of cure but supplies good protection from heat aging is di-beta naphtyl-p-phenylenediamine.

3. ULTRAVIOLET LIGHT INDUCED CROSSLINKING

Because PE contains very few ultraviolet light absorbing groups Oster et al[3,4] discovered that they could decrease the time of irradiation with light in the 200-300 mμ range by a factor of 1000 using benzophenone as a sensitizer to obtain the same degree of crosslinking as without any sensitizer present. The PE film could be allowed to soak in a solution of the sensitizer or the film could be placed along with the benzophenone in a desiccator for two days. It was possible to produce a gel fraction by the UV irradiation in the PE equal to 50-60% and in some cases a value of 85% gel could be obtained. Using a low density PE a gel fraction of 55% could be obtained after 3 minutes of UV irradiation. The UV light does not penetrate into the PE very far, hence it can be used effectively for surface grafting. Using both PE and natural rubber the amount of graft produced with a given incident density of ultraviolet light increased enormously when sensitizers were present in the plastic being irradiated. Because significant extents of grafting could be obtained with very short light exposures no appreciable polymerization of the monomer occurred in the absence of the base plastic. These authors described a simple experiment in which a drop of distilled styrene was placed on the surface of some PE sensitized with 0.25% of benzophenone and covered with a quartz microscope slide (quartz transmits UV light). The sample was then irradiated through the slide, washed with benzene and dried. The graft could be detected because of the ability of the surface to adhere to pressure-sensitive tape under load.

Oster found that both high and low density polyethylenes with the same sensitizer content are equally capable of being crosslinked by far ultraviolot light. Ho pointod out that on tho other hand, however, with ionizing radiation the doses required for crosslinking are greater for high density PE than for low density material. This brings us to an historical description of the radiation crosslinking of polyolefins.

4. HIGH ENERGY RADIATION INDUCED CROSSLINKING

Fairly early in the history of polyethylene as mentioned above it was discovered[18] in 1948 that exposing PE films at room temperature to the high energy radiations of the heavy water pile of the Argonne National Laboratory produced chemical crosslinks, that is C-C covalent bonds between the long chains. This conclusion was deduced from the marked reduction of the cold-drawing ability of the low density PE at room temperature. In 1952 Charlesby[19] also demonstrated from measurements of the reduction in solubility of the PE on pile irradiation that these chemical crosslink bonds had been formed. As explained above, the presence of free radical groups on the PE chains can lead to crosslinks by the recombination of the free radicals on adjacent chains. The mechanisms by which the free radicals can get into positions favorable for the free radicals to recombine will be discussed below. In the meantime, let us consider first how the concentration of crosslinks can be measured or inferred.

Inasmuch as one hydrogen atom must be removed from the hydrocarbon chain in PE to produce a free radical and as it takes the recombination of two free radicals on neighboring chains to form a crosslink, material balance requires that one hydrogen molecule should be produced for each crosslink. However, measurements of the amount of hydrogen gas produced cannot directly be related to the yield of crosslinks because t-vinylene double bonds, -CH=CH-, and a small amount of diene groups are also produced by the elimination of hydrogen during the irradiation. If the concentrations of the unsaturated groups are accurately measured by infrared absorption techniques, then the yield of hydrogen giving rise to the double bonds can be calculated and by difference the hydrogen coming from the crosslink formation can be estimated. This was done by Kang, Saito, and Dole[20] in 1967.

Another method of estimating the yield of crosslinks, or G(X) (in units of number of crosslinks per 100 e.v. of radiation energy absorbed) is to make use of the theoretical equations relating the gel or insoluble fraction of the irradiated sample to the number of crosslinks; see, for example, the theoretical equations of Charlesby and Pinner[21] and of Saito et al.[22] This was done by Kang et all[20] and they obtained a good agreement between G(X) estimated by both methods.

Many other polyolefins have been radiation crosslinked, but as pointed out in 1954 by Miller, Lawton, and Balwit[23] of the General Electric Company (where much of the early studies of the electron beam radiation crosslinking of PE was carried out) polymers of the structure

$$(-CH_2\overset{\overset{\displaystyle CH_3}{|}}{\underset{\underset{\displaystyle R}{|}}{C}}-)_n$$

will degrade on irradiation while those of the structure $(-CH_2CH_2-)_n$

or
$$(-CH_2-\underset{\underset{\displaystyle R}{|}}{CH}-)_n$$

will crosslink. Bohm et al[24] give the following table, Table I, of polymers undergoing either crosslinking or degradation as the major effect.

Table I

Crosslinkable Polymers	Polymers Undergoing Degradation
Polyethylene	Polyisobutylene
Polypropylene	Butyl rubber
Polystyrene	Poly α-methylstyrene
Polyvinylchloride	Polyvinylidene chloride
Polyvinyl alcohol	Polymethacrylates
Polyacrylates	Polymethacryl amides
Polyacryl amides	Polytetrafluoroethylene
Polysiloxanes	Cellulose and its derivatives
Natural rubber	
Synthetic carbon chain rubbers (except butyl rubber)	
Phenol-formaldehyde resins	
Amino-formaldehyde resins	

It should be pointed out that most polymers undergo both crosslinking and chain degradation on irradiation, but there are some for which no degradation has been detected. For example, Pearson et al[25] and more recently Basheer and Dole[26] demonstrated that only crosslinking occurs on irradiating polybutadiene. Basheer and Dole also found the same to be true for block copolymers of butadiene and styrene containing 21 and 61 wt.% styrene.

Katzer and Heusinger[27] radiolyzed polyisoprenes with double bonds in side chains of the type

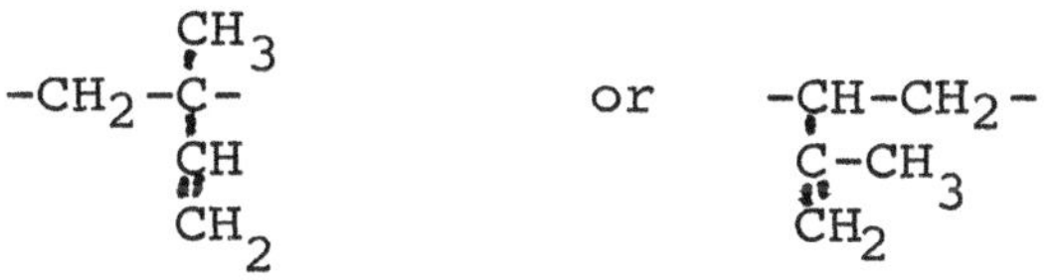

They found that these polymers underwent crosslinking by means of a chain reaction. The concentration of double bonds decreased about 13 times more rapidly than expected. This was explained in terms of intramolecular cyclization according to the reaction involving a positively charged group.

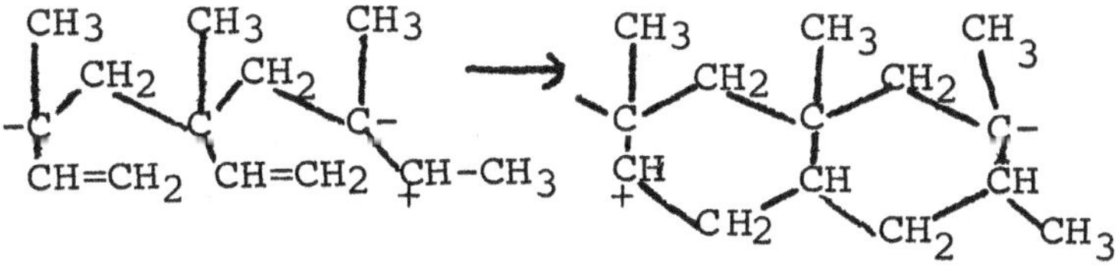
(10)

In the case of the ordinary 1,4 polyisoprenes the crosslinking followed the expected kinetics; that is, one crosslink was produced per initial ionization or excitation.

Crosslinking can be increased by the addition of promoters. For example, Böhm and Tveekrem[28] give a table showing the effect of different types of maleimides on the G(X) values for natural rubber. In 1962 Miller, Roberts and Vale[29] found that the G(X) value increased from 1.9 to 120 by the addition of 5 wt.% each of m-phenylene dimaleimide and dimethyl formamide to the rubber. Very recently Bowmer et al[30] describe the use of trimethylolpropanetrimethacrylate mixed with some diundecyl phthalate as crosslinking promoters in the electron beam irradiation of poly(vinyl chloride). With the TMPTMA in the PVC the dose to the gel point was less than 0.1 Mrad whereas with no TMPTMA in the PVC there was no significant gel formation up to 10 Mrad. Pinner[31] and Miller[32] in 1959 were the first to show that unsaturated monomers could act as crosslinking promoters in the irradiation of PVC.

In conclusion it is interesting to consider possible mechanisms by which the C-C covalent bonds between the long polyethylene chains can form in the solid at room temperature. As is well known free radicals that are produced in polymers by irradiation at 77K are trapped in the solid and cannot react at that temperature even in the case of <u>cis</u> -1.4-polybutadiene that has a very low glass transition temperature,[33] -102°C. On heating PE, for example, to higher temperatures the free radicals decay, but not by a pure second order reaction mechanism as would be expected for the recombination reaction of two free radicals. Instead the alkyl radicals decay by a process[34] which is first order in dose but not in time. French and Willard[35] have called such a process a "composite" first order decay reaction. Some free radicals recombine rapidly as PE is heated to room temperature from 77K, but at room temperature the reaction mechanism probably involves the migration of free radicals across the gap between two parallel chains (the hydrogen "hopping" mechanism) until they get into positions where they can recombine as suggested in 1954 by Dole et al.[36] Evidence for this is the fact that hydrogen gas markedly accelerates the free radical decay as discovered in 1962 by Cracco et al.[37] Presumably the hydrogen catalytic effect is due to the reaction

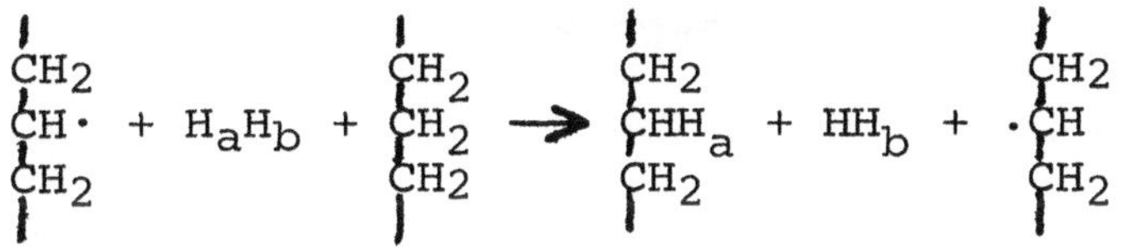 (11)

Proof of this mechanism was obtained by Dole and Cracco[38] who irradiated some PE at 77K, then evacuated the hydrogen produced by the irradiation and at room temperature introduced D_2 gas into the sample cell. HD was produced by reaction 11 above. The rate at which the deuterium exchange decreased with time was quantitatively proportional to the rate at which the free radical concentration decreased. This fact confirms the reaction mechanism given in eq. 11.

ACKNOWLEDGMENTS

Help in preparing this review from Drs. G.G.A. Böhm and B.J. Lyons is gratefully acknowledged. Also the author wishes to express his appreciation of the privilege of using the chemical libraries of the following universities: Harvard, Northwestern, Baylor, and San Jose State, without which privilege this review could not have been written.

Los Gatos Meadows, M-1
Los Gatos, CA 95030

REFERENCES

1. Private communication from G. G. A. Bohm.

2. For a review see M. Dole, J. Macromol. Sci. Chem. 1981, A 15 (7), 1403; 'History of Polymer Science and Technology,' R. B. Seymour, Ed. Marcel Dekker, N.Y. 1983, p. 339

3. G. Oster, J. Polym. Sci. Letters Ed., 1956, 22, 185

4. G. Oster, G. K. Oster, H. Moroson, J. Polym. Sci. 1959, 34, 671

5. B. M. E. Van Der Hoff, Ind. and Eng. Chem., Prod. Res. and Devel. 1963, 2, 273

6. P. Kovacic, R. W. Hein, J. Am. Chem. Soc. 1959, 81, 1190

7. J. H. Drabble, Rubber in Malaga, Oxford University Press, London, 1973, p. 1

8. D. A. Hills, 'Heat Transfer and Vulcanization of Rubber,' Elsevier Publishing Co., Ltd., London, 1971, p. 5

9. R. W. Lunn, 'History of Rubber Industry,' P. Schidrowitz, T. R. Dawson, Eds. W. Heffer and Sons, Cambridge, 1952

10. W. R. Sorenson, T. W. Campbell, 'Preparative Methods of Polymer Chemistry,' Interscience, N.Y. 1961, p. 27

11. L. O. Amberg, 'Vulcanization of Elastomers,' G. Alliger, I. J. Sjothun, L. O. Amberg, Eds., Reinhold, N.Y. 1964 p. 286

12. L. O. Amberg, W. D. Willis, Rubber Age, 1955, 77, 83

13. D. J. Patterson, J. L. Koenig, ACS Symposium Series, 1984, 243, 205

14. R. Basheer, M. Dole, Makromol, Chem. 1982, 183, 2141

15. U. S. Patent, 3,079,370 Feb. 26, 1963

16. H. G. Scott, J. F. Humphries, Modern Plastics, March 1973, p. 82; U.S. Patent 3,646,155 Feb. 29, 1972

17. B. C. Carlson, Rubber World, June 1960, p. 91

18. M. Dole, D. Rose, Chem. and Eng. News, 1948, 26, 2289. Paper presented at the Portland, Oregon meeting of the Am. Chem. Soc., Sept. 1948

19. A. Charlesby, Proc. Roy. Soc. (London) 1952, A 215, 187

20. H. Y. Kang, O. Saito, M. Dole, J. Am. Chem. Soc. 1967, 89, 1980

21. A. Charlesby, S. H. Pinner, Proc. Roy. Soc. (London), 1959, A 249, 367

22. O. Saito, H. Y. Kang and M. Dole, J. Chem. Phys. 1967, 46, 3607

23. A. A. Miller, E. J. Lawton and J. S. Balwit, J. Polym. Sci. 1954, 14, 503

24. G. G. A. Bohm, W. F. Oliver and D. M. Pearson, SPE Journal, 1971, 27, 21

25. D. S. Pearson, B. J. Skutnik and G. G. A. Bohm, J. Polym. Sci, Polym. Phys. Ed. 1974, 12, 925.

26. R. Basheer, M. Dole, Makromol. Chem. 1982, 183, 2141

27. H. Katzer, H. Heusinger, Makromol. Chem. 1973, 163, 195

28. G. G. A. Bohm, J. O. Tveekrem, Rubber Chem. and Tech. 1982, 55, 575

29. S. M. Miller, R. Roberts, R. L. Vale, J. Polym. Sci. 1962, 58, 737

30. T. N. Bowmer, M. Y. Hellman, G. N. Taylor, W. I. Vroom, Org. Coatings and Appl. Polym. Sci. Proceedings, 1983, 48, 496

31. S. H. Pinner, Nature, 1959, 183, 1108

32. A. A. Miller, Ind. Eng. Chem., 1959, 51, 1271

33. R. Basheer, M. Dole, J. Polym. Sci., Polym. Phys. Ed. 1983, 21, 111

34. D. R. Johnson, W. Y. Wen, M. Dole, J. Phys. Chem., 1973, 77, 2174

35. W. G. French, J. E. Willard, J. Phys. Chem., 1968, 72, 4604

36. M. Dole, C. D. Keeling, D. G. Rose, J. Amer. Chem. Soc., 1954, 76, 4304

37. F. Cracco, A. J. Arvia, M. Dole, J. Chem. Phys. 1962, 37, 2449

38. M. Dole, F. Cracco, J. Amer. Chem. Soc. 1961, 83, 2584

E. J. Vandenberg

EARLY POLYOLEFIN SYNTHESIS WITH ZIEGLER CATALYSIS

INTRODUCTION

This paper covers the author's early work at the Hercules Research Center in Wilmington, Delaware, on polyolefin synthesis with Ziegler catalysts. This work began shortly after Hercules Incorporated acquired a license in 1954 to Ziegler's new organometallic-transition metal compound catalysts for polymerizing ethylene at low pressure and temperature to very high molecular weight, linear polyethylene. The emphasis will be on the author's independent discoveries with Ziegler catalysts which led to important commercial developments including (1) isotactic polypropylene, (2) high rate, high yield $TiCl_3$ catalysts for making isotactic polypropylene and (3) the hydrogen method of controlling molecular weight with olefin monomers.

In addition the author will review some pre-Ziegler work on ethylene polymerization, which has only recently been discussed[1], and which yielded linear polyethylene, albeit of somewhat low molecular weight, before it could be easily synthesized, although recognized as a useful entity by an early DuPont patent[2]. This work is of historical interest since a concept of the author's which followed from these studies could have resulted in a logical development of Ziegler catalysis. By one of the common quirks of science, the author did not pursue this early concept. Other investigators missed the Ziegler catalyst discovery in other ways[3].

The Ziegler catalyst was, of course, an unprecedented break-through in the area of polymer synthesis and, even more broadly, in the area of organic reactions. The initial result consisted of the preparation of linear, high molecular weight, high density polyethylene under extremely mild conditions compared to the prior high pressure and high temperature methods required for the commercial, branched, low density polyethylene. This result was startling and led to extensive investigations by many investigators, including

R. B. Seymour and T. Cheng (eds.), History of Polyolefins, 87–101.

the author, to uncover the mechanism and ascertain the scope of what these catalysts could do. Thus, there were unprecedented and feverish research activities for many years and these proved to be very fruitful. The author's contributions were previously reviewed[4] and are repeated here by the courtesy of Plenum Publishing Corp. The early discovery and subsequent development of Ziegler catalysts has been extensively covered by McMillan, including some of the details of the author's work[5].

The Ziegler discovery has thus had wide impact on polymer synthesis as well as on transition metal catalyzed organic reactions - leading to a broad reawakening in the field of inorganic chemistry, particularly on the unique aspects and utility of organometallics. Because of the great importance of this area, the author reviews below some of his other early studies which may have some historical significance.

EARLY POLYMER STUDIES

In 1947, I discovered some very unusual redox systems for emulsion polymerization which consisted of organic hydroperoxides such as cumene hydroperoxide in combination with a reducing agent such as a reducing sugar and a trace of an iron compound[6]. Polymerizations could be done in 15 minutes which previously took 15 hours to do. Hercules was involved in selling a special disproportionated rosin soap for the emulsion polymerization of butadiene and styrene. These new redox systems were much less affected by adventitious impurities and inhibitors. Indeed, one could even use ordinary rosin soaps which contained a variety of conjugated resin acids which inhibited and/or retarded the ordinary persulfate type systems. We filed for patents on these new highly active redox systems and were subsequently awarded patent coverage[7,8]. We speculated at the time that this unusual behavior was due to two factors. First, as shown in Figure 1, the asymmetry of the hydroperoxide causes it to react with a reducing agent to reduce just the OH half of the molecule to OH^- and give an RO• radical. Second, this RO• radical is apparently fairly unreactive so that it does not react with impurities as readily as say a persulfate radical, hydroxyl radical, or perhaps some other peroxide radicals such as from benzoyl peroxide.

$$ROOH + Fe^{++} \longrightarrow RO\cdot + OH^{-} + Fe^{+++}$$

$$Fe^{+++} + \text{Reducing Sugar} \longrightarrow Fe^{++}$$

Fig. 1. Hydroperoxide Redox Systems

Independently, Kolthoff and Medalia at the University of Minnesota[9] and Fryling, et al. at Phillips Petroleum[10] contributed importantly to these new redox systems. The Phillips workers, of course, had a strong commercial interest in this area of polymerization and recognized the value of this to butadiene-styrene polymerization. These higher activity initiating systems permitted them to lower the temperature of polymerization from 50°C to 5°C and this gave a better rubber[12]. Phillips and ultimately other rubber producers adapted these systems for this use. Indeed, organic hydroperoxides have been used and are still being used in making SBR by low temperature emulsion polymerization. Hercules developed and still has a substantial business selling organic hydroperoxides to the emulsion polymerization industry.

In the course of trying to find applications for our new hydroperoxide redox initiating systems, in 1951 we discovered systems which polymerized ethylene at low temperature (5°C) and low pressure (40 atm.), i.e., below the critical temperature (9.9°C). The best system found utilized an insoluble ferrous complex of ethylenediamine-tetraacetic acid with cumene hydroperoxide in 97.5% tert-butyl alcohol (2.5% H_2O), but rates (0.2%/hr.), conversion (2%) and molecular weights (M_w = 8,000) were low. The product, however, was more linear than the commercial high pressure product, since it contained much fewer CH_3 groups by IR, was higher melting (125°C), and was more crystalline by x-ray. In addition, the product was a brittle wax even though we did get the molecular weight up to 20,000 by raising the pressure to 300 atm. This is obviously the linear high density polyethylene which was unknown at that time but which was very easily made a few years later at much higher molecular weight from the Ziegler development and the Phillips chromium oxide on silica catalyst[1]. This polymer has become a very important, large volume commercial product. Our molecular weights were too low for us to recognize its value and, in any event, our process was too poor to be useful. In the course of this work we considered trying a combination of an organometallic such as a Grignard reagent with an organic chloride and a transition metal compound, such as $CoCl_2$, $NiCl_2$, $CrCl_3$, or $FeCl_3$, as a source of free radicals for initiating polymerization. This combination had been reported by Kharasch in his work on free-radical reactions[12], and it falls within the scope of the subsequently discovered Ziegler catalyst. But, the organometallic was not readily available, and we never tried it. So we missed what might have been a logical development of Ziegler catalysis. Indeed, there are examples in the literature where some element of this new coordination polymerization method was used, but in a very imperfect way, and the authors never recognized that it was a new and different method of polymerization[3].

ZIEGLER POLYMERIZATION

General

In mid-1954, Hercules acquired a license to Ziegler's new low pressure ethylene polymerization method which involved a new type of catalyst consisting of organometallics, such as alkyl aluminums, combined with transition metal compounds, such as $TiCl_4$. In October of 1954 I was assigned to do scouting work in this new area of catalysis. Of course, we recognized the importance of Ziegler's discovery. Needless to say, these were very exciting times in the laboratory. We did not know very much about the mechanism of the polymerization method, although it was obvious that some new type of polymerization method was involved. We did speculate that growth of the polymer on the transition metal might occur, although there was considerable debate in the next ten years as to whether the polymer chain was growing on the transition metal or the aluminum metal[3].

Propylene Polymerization

Just one week after I began to work in the field, I tried, on the same day, polymerizing propylene and also using hydrogen in an ethylene polymerization. In the case of propylene I obtained 92 milligrams of an unusual, insoluble, crystalline polymer which I found to be a fiber former. I concluded that it would be attractive for a further study.

In my experiments with hydrogen we ran three experiments plus a control and found that it definitely affected the molecular weight (Table I). The effect was really quite small and I suppose that a lot of people would have considered this inconsequential. However, we did pursue both of these leads.

In the case of propylene, ordinary Ziegler catalysts as described by Ziegler gave very low yields of crystalline polymer; indeed, most of the polymer was an amorphous rubber or even a viscous liquid. Very quickly in 1955, we found catalysts and conditions which gave high yields of the

Table 1. First Experiments on the Effect of Hydrogen on the Polymerization of Ethylene with a Ziegler Catalyst

Mole % H_2	$\frac{\eta_{sp}}{c}$	M_w x 10^{-6}
0	21.0	1.33
0.6	17.4	1.11
1.1	16.4	1.05
2.3	15.8	1.01

crystalline polymer (Figures 2 and 3). Shortly after, Natta and co-workers reported that the polymer was stereoregular and coined the term "isotactic"[13]. We did file a patent application on crystalline polypropylene as a new composition of matter, and were involved in a subsequent interference[1]. However, our dates were not early enough and ultimately we were removed from the interference. In the case of hydrogen, we tried it on propylene and found it to be a very effective way of controlling molecular weight. We filed a patent application which also got involved in an interference. In this case, we prevailed and obtained a patent on our discovery[14]. This method of controlling molecular weight is the major method used in the world today to control the molecular weight of isotactic polypropylene. The production capacity of polypropylene has grown rapidly so that it is now one of the major synthetic polymers with a worldwide capacity of about 14 billion lbs./yr[15].

Catalysts Giving High Stereoregularity - $TiCl_3 \cdot nAlCl_3$. In the case of propylene polymerization, our new catalysts utilized crystalline $TiCl_3$ solids which contained large amounts of $AlCl_3$, up to one mole per Ti. These $TiCl_3$ solids were obviously mixed crystals of $TiCl_3$ and $AlCl_3$ since the x-ray pattern gave no evidence for $AlCl_3$ per se. Both the $TiCl_3$-1.0 $AlCl_3$ and the $TiCl_3$-0.33 $AlCl_3$ mixed crystals gave new and different x-ray patterns which appeared related to the known purple α-$TiCl_3$ rather than the known brown β-$TiCl_3$. Color varied from reddish purple to violet. The $TiCl_3$-1.0 $AlCl_3$ product from $EtAlCl_2$-$TiCl_4$

$$4EtAlCl_2 + TiCl_4 \xrightarrow[\text{48 hrs.}]{25°C} TiCl_3 \cdot AlCl_3 + 3EtAlCl_2$$

Added to C_3H_6 + n-heptane (43 psig initial pressure) + 3 Et_3Al 20 hrs. at 30°C.

$$TiCl_3 \cdot AlCl_3 \quad + 6\, Et_2AlCl$$

"Purple Substance"
δ Form

-95% yield, 86% conv. of heptane-insoluble polypropylene,

$$\frac{\eta_{sp}}{c} = 29$$

Fig. 2 First High Yield Method for Isotactic Polypropylene[16,17]

$$0.33\ R_3Al + TiCl_4 \xrightarrow[90^{\circ}\,C]{16\ hrs.} TiCl_3 \cdot 0.33\ AlCl_3$$

R = Et, IBu

Added to C_3H_6,
n-heptane + 2-4 Et_2AlCl
20 hrs., 30°C.

99% yield, 90% conv. of heptane-insoluble polypropylene,

$$\frac{\eta_{sp}}{c} = 29$$

Fig. 3 Second High Yield Method for Isotactic Polypropylene[20,21] - Stoichiometric Catalyst

reaction was called "purple substance" by our x-ray specialist to differentiate it from the known $TiCl_3$. This type proved important in the commercial manufacture of polypropylene. We initially made this new form by the room temperature reaction of $TiCl_4$ with an excess (2-4 mole/Ti) of $EtAlCl_2$. By using this pre-reaction product with triethyl aluminum in certain ratios where the triethyl aluminum reacted with the excess ethyl aluminum dichloride to form diethyl aluminum chloride (Figure 2), we obtained high yields of stereoregular polypropylene[16,17]. This was our first high yield method of making isotactic polypropylene and may indeed be the first high rate, high yield method discovered by anyone. This method did involve the new so-called "purple substance" form of $TiCl_3$. Hercules obtained a composition of matter patent on this form[18]. Years later, Natta and co-workers[19] reported their excellent crystallographic studies of the various modifications of $TiCl_3$ and called this form δ. It turns out that this δ form was the basis of all the early methods of making polypropylene commercially[15].

Our second and even higher yield (99%) and higher rate method of making isotactic polypropylene was discovered in 1955 as a logical development from our first high yield catalyst. If $TiCl_3 \cdot 1.0\ AlCl_3$ gave this excellent result, how would lower levels of $AlCl_3$ work? We prepared such a lower $AlCl_3$ catalyst by the stoichiometric reaction of trialkyl aluminum with $TiCl_4$, i.e., one alkyl aluminum bond per $TiCl_4$ to form $TiCl_3$-0.33 $AlCl_3$ as the sole product and then using this crystalline solid with Et_2AlCl as an activator (Figure 3)[20,21]. This $TiCl_3$ solid was also a mixed crystal of $TiCl_3$ and $AlCl_3$ (no x-ray evidence for $AlCl_3$) and had a different x-ray pattern from the known "violet" α-$TiCl_3$ but is identical to that finally reported as the γ form by Natta[19], especially when R = i-Bu in Figure 3. This method for making isotactic polypropylene was very effective but the physical form of the product was less favorable and thus was not developed commercially by Hercules. Similar processes have been described in the patent literature[22].

However, this stoichiometric type of $TiCl_3$ has become of major importance in the commercial manufacture of polypropylene[3], based on the reduction of $TiCl_4$ by aluminum metal, as first reported by Ruff and Neumann[23]. Improved methods of making this type of $TiCl_3$ were found and developed by Exxon[24], Stauffer[25], and others[3]. In particular, Tornqvist et al.[26] found that ballmilling was necessary for good polymerization rates. In addition, ballmilling converted the $TiCl_3$ to the δ form[20,27]. Also, the polypropylene was obtained in better physical form. Tornqvist et al.[26-29] also recognized the higher activity and other desirable aspects of the $TiCl_3 \cdot nAlCl_3$ mixed crystals, particularly $TiCl_3 \cdot 0.33\ AlCl_3$, over $TiCl_3$, and ascribed this to a specific effect of $AlCl_3$. This was my belief too. This conclusion is somewhat suspect today since Solvay has reported $TiCl_3$ catalysts of four fold higher activity and with higher stereospecificity than with prior $TiCl_3 \cdot nAlCl_3$ catalysts using especially pure δ-type $TiCl_3$, prepared by special ether extraction and $TiCl_4$ treatment of a Et_2AlCl-$TiCl_4$ reaction product[30]. We were not able to publish, except by patent, in this area of great commercial importance.

Commercial Development Aspects. We went on to make large quantities of isotactic polypropylene and found that it had an interesting combination of properties, although it was not immediately taken up by Hercules since much of our effort was directed toward developing a Ziegler-type process for high density polyethylene. I personally felt that polypropylene would ultimately be an important polymer based on the low cost of propylene. Finally, after about a year or two, Hercules developed a commercial process in collaboration with Hoechst from our lead[31]. As you know, Hercules in late 1957 became the first U.S. producer and is presently the world's largest producer of polypropylene.

This new Ziegler method of polymerization has proved to be very useful commercially and is used on a very large scale for olefin (ethylene and propylene) and diolefin (butadiene and isoprene) polymerization. There has been a tremendous amount of work done on olefin and diolefin catalysts from both the practical and mechanistic point of view and yet we really do not understand the intimate details of this new type of polymerization very well. However, this area is still of great interest. There is much to be learned from a mechanism point of view and there are new products and improved processes of importance to be discovered and developed.

Other Olefin Polymerizations

Our efforts on Ziegler catalysis were directed toward assessing what could be done with this unusual catalysis for (1) overcoming some of the obvious problems for polyethylene manufacture such as molecular weight control (as typified by our discovery of the hydrogen chain transfer method described earlier) (2) studying higher olefin homo and copolymerizations and (3) studying other monomer types. We covered many possibilities by running many experiments per day by utilizing pressure bottles capped with self-sealing rubber liners. We had adopted this technique in the late forties from the emulsion polymerization procedures introduced by rubber polymerization chemists[21]. This technique proved immensely helpful in running many experiments and handling reactive organometallics and other catalyst ingredients under low pressure (< 50 psig) in the absence of air. This work led quickly to three basic patents[14,16,17] which were filed within 9 months from the time we began to work in the field. The earliest patent filed[16] was basic to the subsequently-developed Hercules commercial process for isotactic polypropylene. Another patent[17] covered what may have been the first high yield, high rate process for isotactic polypropylene, as discussed above. The third patent[14] covered the very important hydrogen method for controlling molecular weight. Table II lists some of the important areas that we tested early, as listed in these three patents.

Table II New Areas Tested in Three Early Patents

Area	Patent Reference No.	Example
Homopolymerization of:		
Butene-1	16	45
Butene-1	17	24
Octene-1	16	43
Octene-1	17	22
Octene-1	14	25
Styrene-1	16	46
Butadiene	16	42
Copolymerization of:[a]		
Ethylene with:		
propylene (15%)	17	23
cis-butene-2 (1.6)	16	56
octene-1 (5.8)	16	53
styrene (3.3)	16	47
butadiene (33)	16	57
isoprene	16	54
cyclopentadiene	16	55
Propylene with:		
Styrene (25)	16	58
Butadiene	16	60
Isoprene (25)	14	28
Octene-1 with:		
Styrene (18)	14	27
Butadiene (81)	16	61

[a]% of co-monomer in product given in parenthesis

In these early studies, the homopolymerization of butene-1 and styrene gave unusual crystalline polymers which, as with propylene, were subsequently established as isotactic[13]. Little additional work was done on butene-1 polymerization since isotactic polypropylene appeared more attractive - being higher melting and potentially of lower cost. Isotactic poly(butene-1) was subsequently developed by others as a low volume specialty polymer[1]. Isotactic polystyrene was ultimately studied in some detail. A fairly good, high yield, large sample preparation method was developed[32] and multi-pound quantities of controlled molecular weights were made for applications work. The polymer did not crystallize from the melt unless oriented while cooling. Thus crystalline fibers could be made but ordinary injection-molded, plastic articles in the crystalline state were not feasible. This polymer has not been commercialized.

Octene-1 gave a very tacky, amorphous homopolymer rubber which did not crystallize at room temperature - although presumably it was at least isotactic in part. The melting point of the isotactic polymer is probably below room temperature. This polymer has not been developed commercially.

Butadiene was polymerized but the results were not sufficiently unusual to merit further study, particularly since Hercules had no commercial interest in diene polymers and, also, others quickly made cis-polybutadiene and cis-polyisoprene (synthetic natural rubber) with Ziegler catalysts[3].

In the copolymerization area, it was shown that ethylene could be copolymerized with cis-butene-2, styrene, and dienes; and propylene and octene-1 were copolymerized with styrene and dienes (Table II).

The copolymerization of ethylene with other 1-olefins showed that one could modify the melting point and behavior of linear polyethylene by copolymerization. The author did not study these in great detail. Hercules , Hoechst, and others made butene-1 copolymers with Ziegler catalysts to modify physical properties[33]. It is of special interest that our ethylene-octene-1 copolymer with 5.8% octene-1 falls in the composition range of the so-called linear, low density polyethylene which has very advantageous physical properties and has been developed widely[34]. Subsequently, such materials were covered by a DuPont composition of matter patent[35].

Ethylene-propylene copolymers with propylene contents of 20-80% were subsequently studied by Hercules and others as superior elastomers[1]. Ultimately terpolymers of ethylene, propylene and specific non-conjugated dienes (EPDM) were developed as sulfur-vulcanizable elastomers and are produced commercially in large volume by a number of companies as specialty elastomers[1].

The author also studied polymerizing polar vinyl monomers and cyclic ethers with Ziegler catalysts. This work, although not the subject of this article, led to the discovery of non-Ziegler organometallic coordination catalysts which were more useful for these monomers and led to some interesting new polymers, some of which, as the epichlorohydrin elastomers, were commercialized[4].

Department of Chemistry
Arizona State University
Tempe, Arizona 85287

REFERENCES

1. E. J. Vandenberg and B. C. Repka, in "Ziegler-Type Polymerizations, in "Polymerization Processes, C. E. Schildknecht, Ed., Wiley, New York, 1977, Chapter 11.

2. A. W. Larcher and D. C. Pease (to E. I. duPont de Neumours & Co.), U. S. Patent 2,816,883, filed Aug. 2, 1951, issued Dec. 17, 1957.

3. J. Boor, Jr., Ziegler-Natta Catalysts and Polymerizations, Academic Press, Inc., New York, 1979.

4. E. J. Vandenberg in Coordination Polymerization, C. C. Price and E. J. Vandenberg, Eds., Plenum Publishing Corp., New York, 1983, pp. 13-19

5. F. M. McMillan, The Chain Straighteners, Macmillan Press Ltd, London, 1979.

6. E. J. Vandenberg and G. E. Hulse, Ind. Eng. Chem. **40**, 932 (1948).

7. E. J. Vandenberg (to Hercules Incorporated), U. S. Patent 2, 648, 657-8, filed April 12, 1947, issued August 11, 1953.

8. E. J. Vandenberg (to Hercules Incorporated), U.S. Patent 2,648,655-6, filed November 24, 1948, issued August 11, 1953.

9. I. M. Kolthoff and A. I. Medalia, J. Polym. Sci. **5**, 391 (1950).

10. C. F. Fryling, S. H. Landes, W. M. St. John and C. A. Uraneck, Ind. Eng. Chem. **41**, 986 (1949).

11. W. A. Schulze, W. B. Reynolds, C. F. Fryling, L. R. Sperberg, and J. E. Troyan, India Rubber World **117**, 739 (1945).

12. M. S. Kharasch, F. L. Lambert, and W. H. Urry, J. Org. Chem. **10**, 298 (1945).

13. G. Natta, P. Pino, P. Corradini, F. Danusso, E. Mantica, G. Mazzanti and G. Moraglio, J. Amer. Chem. Soc. **77**, 1708 (1955).

14. E. J. Vandenberg (to Hercules Incorporated), U.S. Patent 3,051,690, filed July 29, 1955, issued August 28, 1962.

15. P. Galli, "Polypropylene - A Quarter of a Century of Increasing Successful Development" presented at IUPAC Macromolecular Symposium, Florence, Italy (September, 1980), Preprints **1**, 10.

16. E. J. Vandenberg (to Hercules Incorporated), U.S. Patent 3,058,963 filed April 7, 1955, issued October 16, 1962.

17. E. J. Vandenberg (to Hercules Incorporated), U.S. Patent 2.954,367, filed July 29, 1955, issued September 27, 1960.

18. E. J. Vandenberg (to Hercules Incorporated), U.S. Patent 3,108,973, filed February 16, 1961, issued October 29, 1963.

19. G. Natta, P. Corradini and G. Allegra, J. Polym. Sci. **51**, 399 (1961).

20. E. J. Vandenberg (to Hercules Incorporated), U.S. Patent 3,261,821, filed December 31, 1959, issued July 19, 1966.

21. E. J. Vandenberg, Isotactic Polypropylene in Macromolecular Syntheses, E. L. Wittbecker, Ed., **5**, 95 (1974).

22. L. W. Gamble, A. W. Langer, Jr., and A. H. Neal (Esso Research and Engineering Company), U.S. Patent 2,951,045 (August 30, 1960); D. E. Winkler and K. Nozaki (Shell Oil Company), U. S. Patent 2,971,925, (February 14, 1961); E. V. Fasce and R. J. Fritz (Esso Research and Engineering Company), U.S. Patent 2,999,086 (September 5, 1961); D. Kaufman and B. H. McMullen (to National Lead Company), U.S. Patent 3,109,822, filed May 28, 1957, issued November 5, 1963.

23. O. Ruff and F. Neuman, Z. Anorg. Chem. **128**, 81 (1923).

24. E. Tornqvist, C. W. Seelbach and A. W. Langer, Jr. (to Esso Research and Engineering Co.), U.S. Patent 3,128,252, filed April 16, 1956, issued April 7, 1964.

25. Preliminary Bulletin No. 56-6, July 9, 1957, and Manual of Procedure for the Evaluation of Olefin Polymerization Catlayst, Ziegler-Natta Type, 2nd ed., Stauffer Chemical Company, 380 Madison Avenue, New York 17, New York.

26. E. G. M. Tornqvist, J. T. Richardson, Z. W. Wilchinsky and R. W. Looney, J. Catal. **8**, 189 (1967).

27. E. G. M. Tornqvist, Ann. N. Y. Acad. Sci., **155**, 447 (1969).

28. Z. W. Wilchinski, R. W. Looney and E. G. M. Tornqvist, J. Catal. **28**, 351 (1973).

29. E. G. M. Tornqvist, Makromol. Chem. **177**, 3437 (1976).

30. J. P. Hermans and P. Henrioulle (to Solvay & Cie), U.S. Patent 4,210,738, filed March 21, 1972, issued July 1, 1980.

31. Hercules Mixer, **41**: 1-10 (January, 1958).

32. E. J. Vandenberg, Isotactic Polystyrene in Macromolecular Syntheses, J. E. Mulvaney, Ed., Wiley, New York, 1977, Vol **6**, 39.

33. W. E. Gloor, Kirk-Othmer Encyclopedia of Chemical Technology, 2nd Edit., Wiley, New York, **14**, 259 (1967).

34. J. N. Short, Kirk-Othmer Encyclopedia of Chemical Technology, 3rd Edit., Wiley, New York, **16**, 385 (1981).

35. A. W. Anderson and G. S. Stamatoff (to E. I. duPont de Nemours and Company), U. S. Patent 4,076,698, filed Jan. 4, 1957 as continuation-in-part of an application filed March 1, 1956, issued Feb. 28, 1978.

J. Paul Hogan and Robert L. Banks

HISTORY OF CRYSTALLINE POLYPROPYLENE

1. BEGINNINGS

An early problem which one encounters in telling the history of a phenomenon or material is that of finding the true beginning. An ancient and reliable document states that "there is no new thing under the sun"[1]. Regardless of what any of us may have discovered, a harbinger or precursor can usually be cited. If one holds a hand mirror before a larger mirror, one can see an ever receding "tunnel" of reflected mirrors, and so it is as we reflect upon the origin of most discoveries.

The history of crystalline polypropylene (PP) is nevertheless unique from several standpoints:

1. Polypropylene was the first high melting crystalline aliphatic hydrocarbon.
2. Polypropylene has been the subject of perhaps the most scientific research and printed matter of any crystalline polymer.
3. Crystalline polypropylene was the subject of the longest, most costly contest, between the most companies, for inventorship of any chemical composition.
4. Crystalline polypropylene is the only hydrocarbon polymer impacting on a Nobel prize.

It thus seems worthwhile of careful and detailed documentation.

Perhaps a reasonable starting point is the discovery of the first polyolefin, semicrystalline (low density) polyethylene in England in 1933-1935[2]. As a result of a happy accident (the right amount of air impurity),ethylene was polymerized to a high polymer at extremely high pressure. However, polymerization of propylene to a high polymer did not occur at similar conditions. Why it would not is fairly clear now from what is known about the effect of a methyl group, but it seemed a very reasonable thing to try in the 1930's. Neither the use of high pressure or any of the catalysts of the 1930's or 1940's resulted in the formation of anything beyond liquids and grease-like materials. Rather high molecular weight propylene polymers were made in the 1940's (with catalysts such as $HBr/AlBr_3$), but they were nevertheless grease-like[3]. Apparently,

R. B. Seymour and T. Cheng (eds.), History of Polyolefins, 103–115.

a carbonium ion mechanism resulted in propylene addition in several modes (not stereospecific), with the result that there was no crystallizable polymer in the product. In fact, the only vinyl polymers known to be crystalline were polyvinyl ethers prepared by Schildknecht[4].

1.1. Early Breakthrough

In the 1940's, Bailey and Reid at Phillips Petroleum Company discovered that nickel oxide on silica-alumina would convert both ethylene and propylene to liquid polymers or oligomers[5]. Hogan and Banks, following up on this discovery in 1949-1951 under Dr. Alfred Clark as Branch Manager and W. C. Lanning as Section Supervisor, found that there were at least 6 different isomers of the dimer of propylene produced over the nickel oxide-silica-alumina catalyst. This fact emphasized the carbonium ion mechanism and lack of specificity[6].

The goal of our work was to produce high octane gasoline fractions by the dimerization-trimerization-etc. of light olefins in cracked gas. One of the problems studied was the regeneration of the fixed bed of catalyst. To attempt to improve catalyst activity and stability during regeneration, chromium (Cr) ion was added as a second transition metal promoter on the NiO-silica-alumina catalyst. On June 5, 1951, propylene dissolved in liquid pentane was pumped over a fixed bed of this modified catalyst at a temperature below 100°C and a pressure sufficient to maintain liquid phase conditions in the reactor. The results were unexpected. In addition to the liquid product normally obtained with the nickel oxide promoted catalyst, a material described in the following record of invention (submitted by Hogan and Banks the same day) was obtained:

> Subject: <u>Synthesis of normally solid polymer from propylene</u>
> Earliest Conception Date: <u>June 5, 1951</u>
>
> "In the polymerization of propylene over commonly used catalysts, such as phosphoric acid and nickel on silica-alumina, the polymer produced is liquid and contains little if any solid polymer.
>
> However, it has been discovered that using a catalyst prepared by impregnating silica-alumina with chromia and nickel oxide, a considerable amount of the product is a white solid, and since a low polymerization temperature is used, this solid is no doubt completely

> olefinic and is therefore a unique solid material. This solid material has a tacky, latex-like nature."

So much has been said and written in the past 30 years about Professor Giulio Natta being the first discoverer of crystalline polypropylene that this non-fact has been almost universally believed in the past. To say otherwise, the authors therefore may seem to be placed in a position similar to that of those brave souls of the 15th Century who dared to assert that the world is round and not flat! Fortunately for the authors, however, a U.S. Federal Court has documented that Hogan and Banks were at least 2 years ahead of Natta in discovering crystalline polypropylene.

1.2. From the Patent Office to the Courts

The legal aspects of the inventorship of crystalline polypropylene appears in detail in another paper. Suffice to say here that following an initial filing of a Phillips patent application (January 27, 1953) and similar applications by other companies in the early 1950's, an interference proceeding was declared by the U.S. Patent Office in 1958[7]. The contestants at first were the authors (Phillips), and scientists at DuPont, Montecatini (later called Montedison), Standard Oil (Indiana), and Hercules. Hercules was dropped from the interference in 1964 because of the late discovery dates at Hercules. Montedison (representing Natta) won the interference in the Patent Office and was awarded a composition of matter patent (U.S. Patent 3,715,344) in 1972, but this decision was appealed by Phillips, DuPont and Standard (Indiana). This resulted in a civil court trial in 1977-78 in Federal District Court (Delaware). The 200-page decision by that Court (January, 1980) awarded priority to Phillips[8]. Appeals by the losing parties resulted in a hearing in the Federal Appeals Court in 1982, and this court upheld the decision of the lower court. The Supreme Court refused to review the decisions of the lower courts, and Hogan and Banks received a patent on crystalline polypropylene on March 15, 1983 (U.S. Patent 4,376,851, assigned to Phillips Petroleum Company). The single claim is identical to the "count" of the earlier interference:

> "Normally solid polypropylene, consisting essentially of recurring propylene units, having a substantial crystalline polypropylene content."

The 200-page OPINION[8] of the U.S. District Court for the District of Delaware, dated January 11, 1980 was based on 14,000 pages of testimony and 4,600 exhibits. Thus, the opinion and the related documents constitute a well supported source of information as to the timing of activities of the various claimants to the discovery of crystalline polypropylene. Following is a condensation of these findings, organized in descending order of priority assigned by the Court[8]. (In the Phillips case, the authors of course relay on themselves as well as on the OPINION.)

1.3. Discovery at Phillips

Following the initial discovery run on June 5, 1951 already described, the next step was to prepare Cr-ion promoted silica-alumina catalyst without nickel ion promoter present. In propylene polymerization runs with this catalyst, none of the liquid product characteristic of the nickel-promoted catalyst was made, and the yield of solid product was improved.

The next important step was to characterize the polypropylene product. Solvent fractionation of the product (first done by Dr. Gene Nowlin) showed that a minor fraction of the total product was insoluble in benzene. The melting point of the first benzene-insoluble fraction was determined by Dr. Nowlin to be 144-151°C. The authors found that this material could be remelted and could be drawn into strands.

Within a few months the first small reactor had been supplemented by a two-inch diameter by four-foot tall reactor which, when operated around the clock, produced several pounds of polypropylene in each run. Solvent fractionation of this material provided sizable quantities of the highest molecular weight, insoluble fractions which were then further characterized by melting point, density (0.91-0.92), and microstructure by infrared.

The same 2-inch I.D. reactor was used to produce, in 1951-1952, several-pound quantities of high density polyethylene and homopolymers of other alpha-olefins including poly(4-methyl-1-pentene). Nowlin had already shown that the poly(4-methyl-1-pentene) fraction insoluble in common solvents was very high in melting point (210-216°C).

Four U.S. patents eventually issued, directed to the subject matter first disclosed in the patent application filed by Hogan and Banks in behalf of Phillips on January 27, 1953. They are:

1. Polymers and Production Thereof, U.S. Patent 2,825,721 (March 4, 1958);
2. Polymerization Catalysts, U.S. Patent 2,951,816 (September 6, 1960);

3. Solid Polymers of 4-Methyl-1-Pentene, U.S. Patent 4,342,854 (August 3, 1982);
4. Polypropylene, U.S. Patent 4,376,851 (March 15, 1983).

1.4. Karl Ziegler and Giulio Natta Discoveries

Dr. Giulio Natta, in a deposition taken in 1970, testified that on March 11, 1954 he made the following diary entry: "Today we polymerized propylene." He further testified that it was his idea to polymerize propylene, and that he shared his idea with his assistants, Pino, Mazzanti, and Chini. Chini made the initial successful run and it was shown by x-ray that it was partly crystalline.

The Natta story begins not in Italy, however, but in Germany. At the Max Planck Institute in Mulheim, W.Germany, Professor Karl Ziegler and his assistants discovered in about 1950 that aluminum alkyl caused the conversion of ethylene to oligomers with an even number of carbon atoms. The next step was the fortuitous discovery that the presence of nickel ion caused the ethylene to be converted almost totally to butenes. A systematic study of the effects of the addition of other metal ions to the aluminum alkyl led, in October of 1953, to the discovery that zirconium ion plus aluminum alkyl polymerized ethylene to a high molecular weight polymer. Within a month after this break-through, titanium ion was tried, and polyethylene was produced at high rate. (Ziegler and coworkers were over two years later than Hogan and Banks in discovering catalytic conversion of ethylene to linear solid polymers.)

In 1952 Natta learned of Ziegler's work with ethylene and triethylene aluminum and immediately initiated an invitation to Ziegler to visit Milan, Italy at Montedison Co.'s expense. (Natta was already under contract to Montedison, then known as Montecatini.) A result of this visit was a contract whereby Montedison purchased certain commercial rights in Italy to Ziegler's discoveries. The agreement also included some scientific exchange of information between Ziegler and Natta. As an immediate result, three young Montedison scientists, already assigned to Natta, were permitted to work at Ziegler's institute. Consequently when the polyethylene breakthrough occurred in late 1953, one of these scientists was present at the Max Planck Institute. This news soon reached Milan, and a letter from a Mr. Orsoni of Montedison to Ziegler, dated December 17, 1953, asked for details of the breakthrough. Ziegler's reply, dated January 6, 1954, stated that copies of two recent patent applications were being sent to Montedison (revealing Ziegler's discovery of titanium halide/aluminum alkyl catalyzed ethylene polymerization to high polymer). In the same letter, Ziegler urged that Montedison "start something practical" in the way of commercial development and leave experimental discovery to Ziegler. Only two days before Natta's production of crystalline polypropylene with Ziegler's titanium tetrachloride/alumina alkyl catalyst, Ziegler visited Milan and restated his position that exploratory research with Ziegler's new catalysts was to be left to the Max Planck Institute.

Natta's testimony in 1970 was that Ziegler had stated that propylene could not be polymerized to high polymers. Ziegler's testimony (1967 deposition) was that an early attempt in December of 1953 was not fruitful but that special polymerization equipment was being built to try again when Natta's first run was made. Ziegler sent a letter to Orsoni July 21, 1954 transmitting a small sample of partly crystalline polypropylene made at the Max Planck Institute, and only then was Ziegler informed that Natta had already succeeded in making polypropylene. Natta, P. Pino, and G. Mazzanti filed U.S. Patent Applications 514,097 and 514,099 claiming a process and crystalline polymers, respectively, of various alpha-olefins including propylene on June 7, 1955. Polypropylene density and m.p. were given as 0.92 and 160°C respectively. They claimed (and received) priority dates of June 8, 1954 and July 27, 1954, the dates of filing of Italian patent applications. These patent applications were in behalf of Montedison (then Montecatini). Ziegler's patent applications were, of course, slightly later than Natta's. The resulting legal conflict between Ziegler and Montedison led to settlements in which Ziegler withdrew from the fight for patent rights on crystalline polypropylene in exchange for a minor share in royalties which Montedison might receive. The possible personal conflict between these two renowned scientists apparently ended a close personal relationship, but they shared a Nobel prize in Chemistry in 1963, recognizing their outstanding contributions to polymer science. In their acceptance lectures, each acknowledged the contributions of the other.

Natta soon discovered that $TiCl_3$ was a more efficient producer of crystalline polypropylene than $TiCl_4$, and that the violet ("delta") crystalline form of $TiCl_3$ was preferred over the others. Natta was the first to use the term "Ziegler catalyst", and "Ziegler-Natta" catalysts soon became the principal basis of a rapidly growing polypropylene industry.

1.5. Standard Oil (Indiana)

In work supervised by B. L. Evering in the Exploratory Research Division at Whiting, Indiana, Alex Zletz, in 1950, used a hydrogen-reduced molybdena on alumina catalyst to polymerize ethylene[8]. Later the same year he made several attempts to polymerize propylene with similar catalysts, apparently without significant success. In the same laboratories, Carmody used reduced cobalt molybdate on alumina in September of 1950 in two or more attempts to polymerize propylene at 170°C. Very small amounts of solid polymer, insoluble in methyl ethyl ketone, was obtained. However, the infrared scannings of these materials revealed that the methylene to methyl ratio was at least 1.6 and possibly as high as 2.6. Since the $-CH_2-/-CH_3$ ratio in a homopolymer of propylene is essentially 1, these samples resembled ethylene-propylene copolymers. (Since this catalyst was later discovered to catalyze olefin metathesis, converting propylene to ethylene and butenes at an optimum temperature of about 160°C[9], the source of the unintended ethylene is apparent.)

In May of 1953, E. F. Peters of Standard (Indiana) used the cobalt-molybdena catalyst of Carmody to attempt the polymerization of propylene at room temperature. After 1 month reaction time, boiling xylene extraction of the catalyst yielded a small amount of polypropylene (precipitated by cooling the xylene) which infrared examination showed to have a much lower $-CH_2-/-CH_3$ ratio than Carmody's polymers and to have some crystallinity. (Since the olefin metathesis reaction would have been largely eliminated at room temperature[9], the reason for the difference in results by Peters and Carmody is apparent. Peters much later became aware of the ability of molybdena-alumina catalyst to "crack" propylene to ethylene and 1-butene[10]. There were apparently no determinations of melting point, density, or solution viscosity of these polymers made in May 1953. Standard Oil (Indiana) Patent Application No. 462,480, filed by Alex Zletz on October 15, 1954, recited a density between 0.85 and 0.95, a softening point above 120°C, and a viscosity above 0.1 for polypropylene.

The experimental work on olefin polymerization with molybdena catalyst eventually led to a commercial process for ethylene polymerization [11].

1.6 DuPont

The first bench-scale production of what was later shown to contain a portion of crystalline polypropylene was the work of Dr. N. G. Merckling of Dupont [8]. Merckling in early 1954 discovered a 1953 German Patent by Max Fischer (BASF) describing the production of white solid (polymer) when ethylene was passed over a catalyst made by heating $AlCl_3$, $TiCl_4$, and Al powder. (Fischer had apparently done this almost 10 years earlier, during WW II.). After producing polyethylene with this catalyst, Merckling modified it to consist essentially of $TiCl_4$ plus a Grignard compound, and by late April of 1954 had produced a small amount of polypropylene. The research project was then expanded to include A. Anderson, G. S. Stamatoff, and W. N. Baxter, and by May 17, 1954 propylene polymers which contained crystalline polypropylene had been made. The melting point was about 150°C and the density was 0.88 g/cc. In a Dupont patent application filed August 19, 1954, by the above 4 scientists, $LiAl(n\text{-}hexyl)_4$ and $Sn(n\text{-}butyl)_4$ as well as Grignard compounds were used with $TiCl_4$ to polymerize propylene to polymers containing a crystalline component.

1.7 Hercules and Others

Early olefin polymerization studies at Hercules Powder Co. were not described in the Court Decision referred to earlier, since Hercules was eliminated from the Intereference in 1964 by the U.S. Patent Office because of their late discovery and patent application date [12]. However, Edwin Vandenberg has told the story [13] and will be relied on here.

Hercules obtained one of the first Ziegler polyethylene process licenses in mid-1954. Vandenberg was given the assignment to do scouting work with the new Ziegler catalysts in October of 1954. Within a week he had polymerized propylene with Ziegler catalyst and had separated out of the product a small fraction of an "unusual, insoluble, crystalline polymer." Catalyst research led, in early 1955, to the development of improved catalysts made by reducing $TiCl_4$ with aluminum alkyls (external to the polymerization reactor) to give $TiCl_3{\cdot}AlCl_3$. Vandenberg discovered the use of hydrogen to control molecular weight of polyolefins made with Ziegler-Natta type catalysts [14], and this remains the principal method of control today.

Crystalline polypropylene was apparently made by others who licensed Ziegler's catalysts, before Natta's public announcement in early 1955 of his discovery. At least two cases have been reported. At both Petrochemicals, Ltd. in England and Hoechst in W. Germany, polypropylene was made in 1954 (in the first case soon after Ziegler and the second case before Ziegler but after Natta). At neither company was there any attempt made to patent what seemed to belong to Ziegler [15].

2. COMMERCIALIZATION

2.1 Industrial Aspects

The commercial development of crystalling polypropylene was rapid and extensive. Initial commercial production started in late 1957, less than four years after public disclosure of this new material. The first commercial producers were Hercules in the United States, Farbwerke Hoechst in Germany, and Montecatini (later Montedison) in Italy [16,17]. Today crystalline polypropylene is produced by many companies in plants throughout the world with world-wide demand exceeding 13 billion pounds in 1983. Total world-wide capacity in 1984 was greater than 18 billion pounds per year, with about 30% of the capacity in the United States [18]. Companies producing polypropylene in the United States in 1981 were Amoco, Arco Polymers, Dart Industries, Exxon, Gulf Oil, Hercules, Northern Petrochemicals, Novamont, Phillips, Shell Chemical, Soltex Polymers (Solvay), and Texas Eastman [16].

Technology for the early polypropylene plants stemmed from polyethylene process technology. In fact, some of the polypropylene plants still in operation today were originally constructed for high density polyethylene manufacture with Ziegler catalysts [19]. The early plants used a slurry process in which an inert liquid hydrocarbon diluent (e.g., hexane, heptane) was used in stirred autoclaves at temperatures below about 90°C and at pressures sufficient to maintain 10-20% propylene monomer concentration in the liquid phase. The traditional catalyst was a crystalline, violet color form of $TiCl_3$ plus diethyl aluminum chloride co-catalyst. The polymer formed as particles and was removed as a 20-40% solid slurry, which was flashed to remove unreacted propylene. Catalyst was deactivated and solubilized by addition of HCl plus alcohol or other polar organic compounds, and removed from the polymer by centrifuging, filtration, or aqueous extraction. Amorphous polypropylene was separated as a solution in the liquid hydrocarbon and could be recovered by evaporation of the solvent. These first plants were somewhat inefficient, with very low catalyst productivity and relatively low crystalline polymer yields (75-85%). The slurry technique is still the most widely practiced commercial process for producing polypropylene, and although major improvements in both catalyst activity and stereospecificity have occurred during the past quarter-century, for some plants expensive catalyst and amorphous polymer removal steps are still necessary.

Solution polymerization, one of the oldest of the commercial polypropylene processes, is generally considered obsolete. the higher temperature (140 to 150°C) necessary to keep the polymer in solution results in the production of greater amounts of normally undesirable amorphous polymer than by other commercial processes. Texas Eastman, who apparently has an internal need for the amorphous polymer, is the only major polypropylene producer in the United States using this process.

A bulk or liquid monomer propylene polymerization process was commercialized by Phillips Petroleum Company in 1963. The process is designed to take advantage of high polymerization rates associated with bulk polymerization and high heat transfer rates characteristic of loop reactors. Liquid propylene (plus co-monomer if desired), catalyst components, and small amounts of hydrogen (for molecular weight control) are fed continuously to the reactor while propylene-polymer slurry is removed. The loop reactor is operated liquid full; propylene-polymer slurry is circulated rapidly by a single pump. Catalyst is removed from the granular polymer by solubilization with polar organic compounds and countercurrent washing with liquid propylene, which also removes a small amount of soluble polymer. The polypropylene is then freed from the propylene in a flash chamber, and is dried and pelletized.

A bulk polymerization system in which the heat of reaction is controlled by vaporization of propylene was developed and used commercially by Dart Industries. In a Rexene version of bulk polymerization, agitation in the three-phase reactor system is maintained by continuous addition of gaseous-olefins to the bottom of the reactor at velocities sufficient for fluidization of solid particles.

A gas-phase process for producing polypropylene was commercialized by BASF in a small plant in West Germany in the mid-1960's. The polymer initially produced had higher than normal amorphous content and low stiffness and hardness properties. In 1972, BASF started producing polypropylene with a much higher isotactic content and improved properties. Northern Petrochemical licensed this process and started production in the United States in the mid-1970's. By 1983 it was claimed that this process had the ability to produce higher crystalline products needing no removal of amorphous polymer and minimal stabilization [20]. Gas-Phase technology has also been developed by Amoco, which operates a large gas-phase polypropylene plant in Texas. Currently, a joint Shell/Union Carbide gas phase process utilizing Shell high activity catalyst and Union Carbide fluid bed technology is being commercialized [21].

2.2 Catalyst Advances

The essential factor in commercial production of highly crystalline polypropylene is obviously the catalyst system. Major catalyst development goals have been to increase both activity and stereospecificity to levels where removal of the catalyst (deashing) and amorphous polymer from the product can be eliminated. For most commercial uses of polypropylene, the first goal requires Ti and Cl residue in the polymer of less than 5 and 20 ppm, respectively; the second goal limits production of the amorphous (soluble) fraction to 4% or less [17]. In pursuit of these goals, five plateaus of commercial catalyst performance have occurred: in situ production catalyst (1957), external production catalyst (1960), promoted catalyst (1970), high surface area catalyst (1976), and supported catalysts (1977) [17].

Preparations of catalyst for the initial commercial plants were by reduction of $TiCl_4$ with alumina alkyl compounds directly in the polymerization reactors. These catalysts were, at least partly, the less active form of $TiCl_3$ (i.e., brown beta form) and had poor stereospecificities requiring both deashing and extensive extraction of the polymer to remove the amorphous fraction (15-25%). Purple $TiCl_3$, which can be obtained by thermal treatment of the beta phase. was found to give high stereospecific product. Commercial quantities of purple $TiCl_3$ from aluminum reduction of $TiCl_4$ was produced in the United States by Stauffer Chemical Company starting in 1960, marking the beginning of marketing of the external production catalyst. However, Hercules apparently produced purple $TiCl_3 \cdot AlCl_3$ external to the reactor by reduction of $TiCl_4$ with $(C_2H_5)_2AlCl$ well before then [13]. These developments were followed by the discovery that intense dry grinding of the catalyst in ball mills increased activity. With the change to external production catalysts, activities were increased by an order-of-magnitude and formation of amorphous polymer decreased to less than 10% of the product.

The next significant catalyst development was the use of special catalyst adjuvants. It was found that electron donors (or Lewis bases) added to the slurry or incorporated into the $TiCl_3$ significantly improved stereospecificities to the level where in many cases atactic removal facilities were not needed. Thousands of electron donors compounds have been claimed to be effective; however, only a few are used commercially. In many cases catalyst activity is adversely affected. An example of one class of electron donors is the monoterpenic ketones developed by Stauffer Chemical Company [17].

The next stage of development addressed eliminating the

catalyst removal step. Procedures to produce higher surface area catalysts were developed by Solvay and Shell International [17]. Compared to previously known catalysts, the high surface area catalysts were two to four times more active and did not adversely affect crystallinity. However, deashing was still required. The first higher surface area catalysts had the problem of stability, requiring on-site production.

The most recent major development is supported titanium catalyst. The goal of the development of supported catalysts was activity and stereospecificity sufficiently high to avoid both catalyst and amorphous polymer removal. In 1976 Montedison and Mitsui disclosed that 300,000 grams of polymer per gram of titanium in combination with crystalline content of 95% had been obtained. In most patent examples, the supported material is anhydrous magnesium chloride or other magnesium compound. Supported Ti catalysts for propylene polymerization are now used in plants in several countries including the U.S. Improvements in supported catalysts and adjuvants have continued to the present, and U.S. patent applications for supported polypropylene catalysts have exceeded 75 per year since 1977 [22].

3. BENEFITS

More than 14 billion pounds of crystalline polypropylene is now being produced world-wide annually, which is equivalent to nearly three pounds per person; in the United States, production approaches 20 pounds per person. Major applications of crystalline polypropylene are injection-molded components, especially for the automotive industry and for electrical applications, fibers and monofilaments, film and sheet, pipe and conduit, and blow molding and other extruded products. Although homopolymer is the principal product, random and block copolymers containing small amounts of ethylene are also commercially important. It is anticipated that the demand for polypropylene will continue to grow for the rest of this decade and new applications requiring specialized-types of polymers will continue to emerge.

4. REFERENCES

1. Ecclesiastes 1:9.
2. E. W. Fawcett, R. O. Gibson, M. W. Perrin, and E. G. Williams (to ICI), British Patent 471,590 (accepted Sept. 6, 1937).

3. C. M. Fontana, R. J. Herold, E. J. Kinney, and R. C. Miller, Ind. Eng. Chem. 44, (12), 2955-2962 (1952).
4. C. E. Schildknecht, et al., Ind. Eng. Chem. 40, 2104 (1948); Ibid 41, 1998, 2891 (1949).
5. G. C. Bailey and J. A. Reid (to Phillips Petroleum Co.), U.S. Patents 2,381,198 (1945); 2,581,228 (1952); 2,606,940 (1952).
6. J. P. Hogan, R. L. Banks, W. C. Lanning, and A. Clark, Ind. Eng. Chem., 47, (4), 752-757 (1955).
7. Interference No. 89,634, initiated by the U.S. Patent and Trademark Office, Sept. 9, 1958.
8. OPINION, dated January 11, 1980, United States District Court for the District of Delaware, Civil Action Suit 4319 (U.S. Patent Quarterly, p. 676).
9. R. L. Banks and G. C. Bailey, Ind. Eng. Chem. Prod. Res. and Dev. 3, (9), 170-173 (1964).
10. E. F. Peters and B. L. Evering (to Standard Oil of Indiana), U.S. Patent 2,963,447 (1960).
11. E. L. d'Ouville in "Polythene", 2nd Edition, A. Renfrew and P. Morgan, eds., Illife & Sons Ltd, London, p. 35-40 (1960).
12. E. J. Vandenberg (to Hercules Inc.) U.S. Patent Application 500,051, filed April 7, 1955.
13. E. J. Vandenburg in "Coordination Polymerization," C. C. Price and E. J. Vandenberg, eds. Plenum Publ. Corp., p. 11-43 (1983).
14. E. J. Vandenberg (to Hercules, Inc.) U.S. Patent 3,051,690, filed July 29, 1955; issued Aug. 28, 1962.
15. F. M. McMillan, "The Chain Straighteners," The Macmillan Press, London, p. 111-112 (1979).
16. H. R. Sailors and J. P. Hogan, J. Macromol. Sci.-Chem., A15, (7), 1377-1401 (1981).
17. K. P. Triplett in "Applied Industrial Catalysis," Vol. 1, B. E. Leach, ed., Academic Press, Inc. p. 177-205 (1983).
18. G. Crespi and L. Luciani in "Kirk-Othmer Encycl. Chem. Technol.," 3rd Ed., M. Grayson and D. Eckroth eds., Vol. 16 pp. 453-69 (1981).
19. J. N. Short Rubber and Plastic News, May 26, 1980, pp 22-23.
20. R. L. Hull, R. H. Foster, and W. A. Bowles, SPE Technical Papers, RETEC Meeting Feb. 27-28, Houston, 1984, p. 239-245.
21. Plastics World, December, 1983, pp. 6-7.
22. R. W. Glass. SPE Technical Papers, RETEC Meeting Feb. 27-28, Houston, 1984, pp. 37-43.

Tai C. Cheng

AN OVERVIEW OF POLYPROPYLENE RESINS

INTRODUCTION

Polypropylene is classified as the youngest of the used plastics. This material has become one of the largest and fastest-growing thermoplastics in the market place today. From its introduction as a commercial plastic in 1975, it has grown to a 2.92 billion-pound-per-year business in 1979. The growth rate of polypropylene resin will reach 4.19 billion pounds per year by 1985. The rapid growth of polypropylene is due to its versatility, wide applicability, and low cost.

Although polypropylene has been studied extensively at Raychem, it has not been used in any of Raychem's products. The reasons for this are that, after irradiation, (i) polypropylene ages poorly, and (ii) polypropylene loses mechanical strength.[1] These problems are believed to be associated with the large number of molecular branch points at the methyl groups bonded along the polymer backbone. Such branch points are sensitive to thermal and oxidative degradation. The fragments resulting from degradation show little tendency to combine with other polymer chains to form long branches; the fragments (tertiary radicals) are more stable than secondary radicals. The order of the stability of radicals is shown as follows:

$$-CH_2-\underset{\bullet}{\overset{\overset{CH_3}{|}}{C}}-CH_2 > -\overset{\overset{CH_3}{|}}{C}-C\,\cdot > H-\underset{\bullet}{\overset{\overset{CH_3}{|}}{C}}-H > CH_3\,.$$

As a result, polypropylene chains are easily broken at high temperatures, normal processing temperatures and during radiation. These problems could be reduced if polypropylene is blended with a polybutadiene, a partially hydrogenated polybutadiene, or other polymers containing double bonds.

R. B. Seymour and T. Cheng (eds.), History of Polyolefins, 117–130.

Although the method of using methacrylate[2] or butadiene[3] as cross-linking agents to improve the mechanical properties of polypropylene has been reported, the information presented in the literature is sketchy and the materials do not contain carbon black. Thus, it is important for us to re-evaluate polypropylene blended with unsaturated polymers. It is our hope that with such binary materials we could also improve the flexibility of polypropylene-based products.

This report is a brief summary of the history, synthesis, properties and products of isotactic polypropylene. Although crystalline polypropylene is produced in two forms (isotactic and syndiotactic), only isotactic polypropylene will be discussed. Syndiotactic polypropylene is omitted due to the lack of commercial interest (low polymerization temperature required, e.g., -78°C). A detailed report regarding the preparation and characterization of syndiotactic polypropylene has been published by Boor.[4]

HISTORY OF POLYPROPYLENE

In 1873, A French chemist, Prunier, obtained a poly-propylene polymer for the first time, but it was a liquid polymer and had no practical use.

In 1954, Professor Giulio Natta and a group of Montedison research workers began work on the polymerization of mono-olefins and di-olefins using Ziegler catalysts. They discovered the stereospecific polymerization of alpha-olefins. That finding led to many important industrial patents. For these reasons, Professor Natta was awarded the Nobel Prize in Chemistry along with Professor Ziegler.

In 1975, Montedison in Italy and Hoechst in West Germany along with several U.S. manufacturers began production of isotactic polypropylene[5,6]. Thus, polypropylene has been a viable thermoplastic since 1958.

SYNTHESIS OF POLYPROPYLENE

Polypropylene is prepared by the reaction of polypropylene in the presence of a catalyst which consists of an ogranometallic compound (e.g., triethyl-aluminum) and a transition metal compound (e.g., titanium tetrachloride) in a dry, oxygen-free, inert hydrocarbon solvent. This catalyst has been referred to as a Ziegler-Natta catalyst. The production process based on Ziegler-Natta systems can be classified as follows:

(a) Slurry Process

The earliest polypropylene plants were built using a slurry process. This process, in general, involves a solvent (hexane or heptane), propylene, catalyst, and a small amount of hydrocarbon in a stirred autoclave at a temperature below 90°C. The propylene concentration of this process is about 10 to 20 percent, depending on the type of catalysts used. The traditional catalyst for the slurry process is $TiCl_3$ plus diethyl aluminum chloride (as a co-catalyst). However, the reduction of $TiCl_4$ with an aluminum alkyl is also used as a catalyst for the slurry process. The final polymer, as particles, is removed from the autoclave and the catalyst in the polymer mixture is deactivated and dissolved by means of alcohol and HCl or other polar organic compounds. Currently, this process is still used by Amoco, Exxon, Gulf, Hercules, Novamont, Shell and Soltex for the production of isotactic polypropylene.

(b) Solution Process

Solution polymerization has been used commercially in the U.S. since 1960. Such a process, in general, produces mainly an atactic polypropylene due to concentration effect. Texas Eastman Company is currently producing an atactic polypropylene using the solution polymerization technique.

(c) Bulk Process

Bulk polymerization was developed by Phillips [7,8]. In this process, liquid propylene is the only hydrocarbon diluent. The residue of the Phillip-catalyst in the polymer mixture is removed by solubilization with polar organic compounds which are washed from the granular polymer by a counter-current stream of liquid propylene. The polymer is then separated from propylene by flashing. Currently, both Phillips and ARCO are using this process for the production of isotactic polypropylenes.

(d) Bulk-Boiling Process

The bulk-boiling polymerization technique was developed by Dart Industries. This system is a modified bulk polymerization technique. Basically, this process enables liquid propylene to boil so that the heat of the stirred reaction is removed in a vapor space at the top. This process is currently used for the production of polypropylene at Dart.

(e) Gas Phase Process

This polymerization technique was developed by BASF. The drawback of this process is the higher amount of amorphous polypropylene as a by-product. This process is currently used by BASF and Northern Petrochemical. Very recently, Amoco has apparently put on stream a large new gas phase plant in Texas based on technology developed by Amoco.

MECHANISM

The exact mechanism of the propylene polymerization is unknown. However, it generally can be divided into two classes based on the solubility of catalysts. One is called homogeneous phase catalysis while the other is called heterogeneous phase catalysis. In homogeneous catalysis, it is clear that the factors controlling the stereospecificity must be linked to specific interaction of the monomer and/or the growing chain with the active complex. This, of course, includes metal, counterion, ligands and solvent. In the heterogeneous case, it is speculated that the structure of the catalytic surface is the determinant factor. The rule of thumb for the Ziegler-Natta systm is to understand the catalyst activity prior to controlling the molecular weight of the final polymer. Otherwise, one might produce a polymer leading to an undesirable product. An example of this case is the isotactic versus the atactic polypropylene. The former material is a plastic with a melting point of 165°C, while the latter product is a rubber with no melting point. Thus, applications of these two polymers are naturally different.

Although the exact role of the organometallic compound is still debatable, it appears to be narrowed down to two mechanisms. One is limited to alkylating the transition metal. The propagation step of this mechanism involves two or more coordination sites on the central transition metal. This is shown as follows:

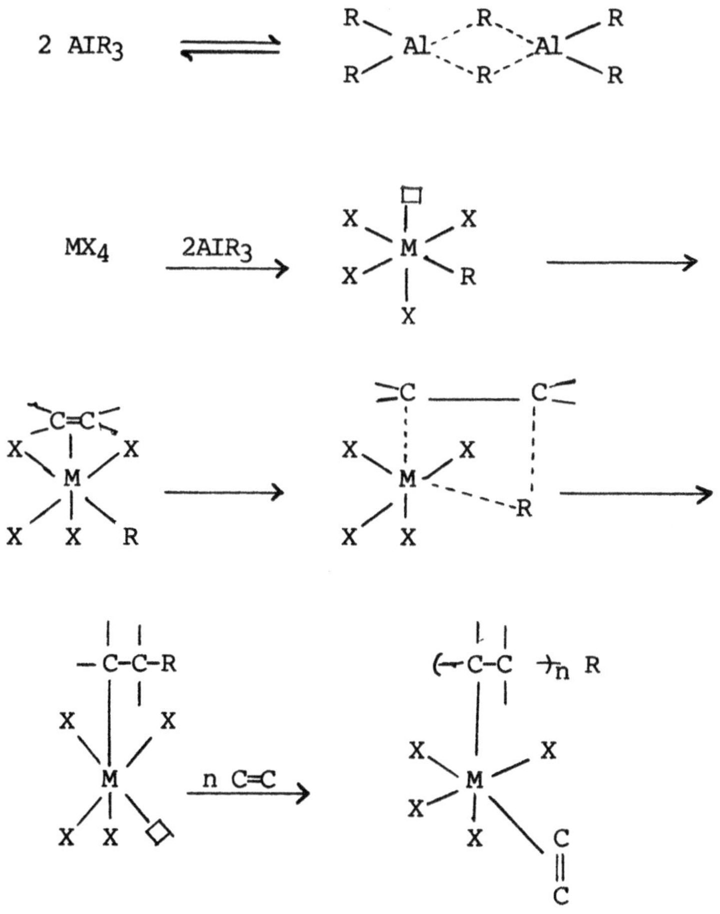

where M is a transition metal, X is a lingand, and □ is a vacant orbital serving as the reaction site. The other possible mechanism for this system has also been accepted by scientists and involves both alkylation and participation of the metal in the formation of an active catalyst center. The detailed reaction mechanism is shown as follows:

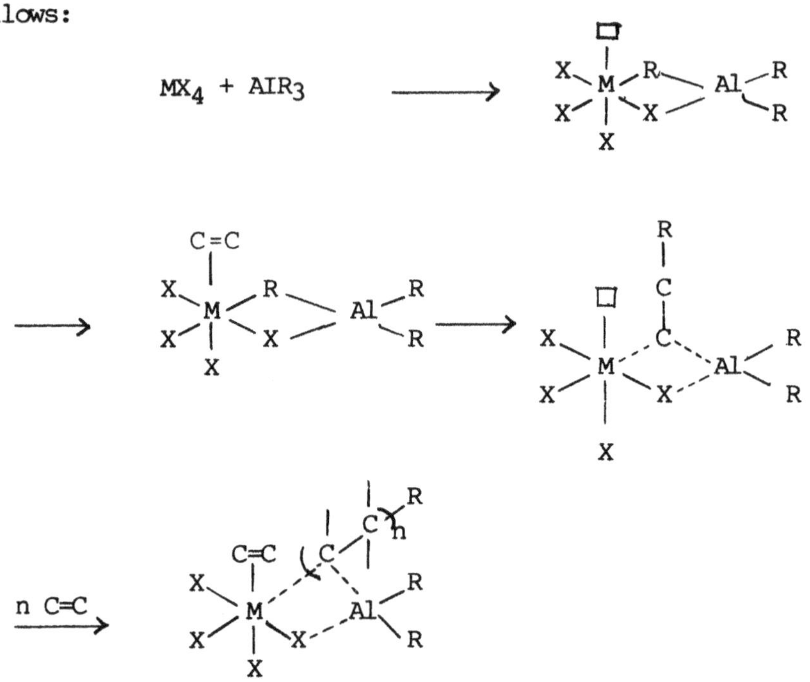

Again, M is a transition metal, X is a ligand, and □ is a vacant orbital serving as the reaction site.

The detailed information regarding the mechanism of the propylene polymerization is published by Saltman[9] and Boor[10] and will not be repeated in this paper.

STRUCTURE AND PROPERTY

The structure of the polypropylene molecules is complex. It can be regarded as a polyethylene chain with a methyl group bonded to every other carbon atom in the polymer backbone. Based on the arrangement of these methyl groups along the molecular chain, one can visualize three basic structures for polypropylene: isotactic, syndiotactic, and atactic polypropylene. The structures of polypropylene are shown as follows:

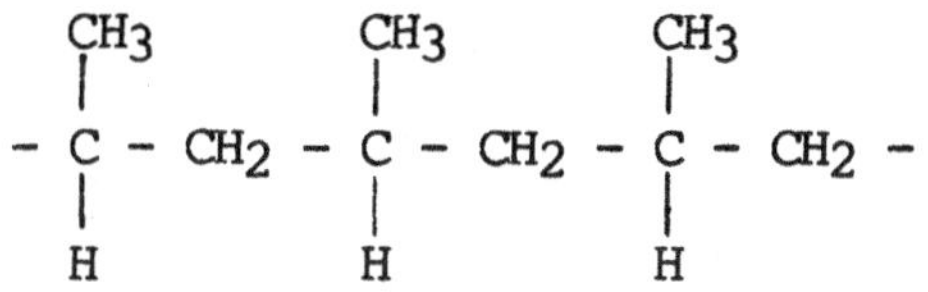

Isotactic (I)

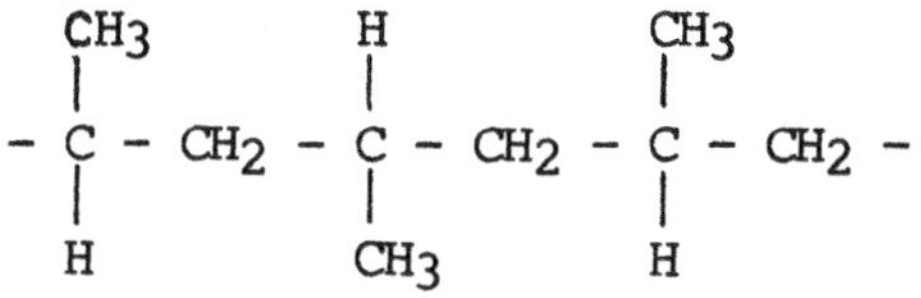

Syndiotactic (II)

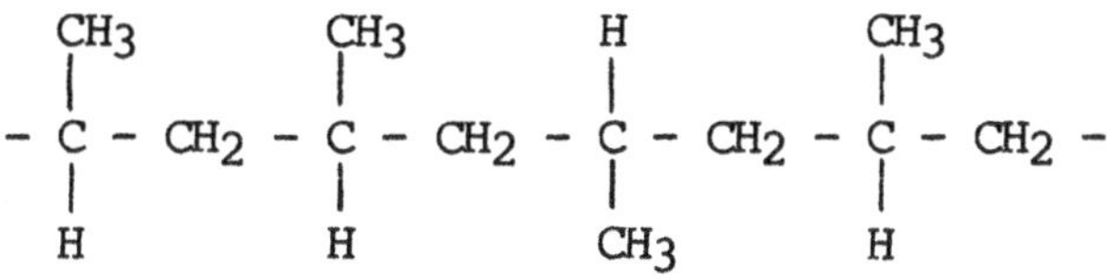

Atactic (III)

Isotactic polypropylene (I) can be described as all methyl groups on one side of the polymer chain, whereas syndiotactic (II) and atactic (III) polypropylene do not have the methyl groups bonded on the same side of the polymer chain. In general, crystalline structure of polypropylene is obtained if the methyl groups are arranged in a regular sequence on one side of the polymer chain. Otherwise, an amorphous material is obtained.

The range of properties of polypropylene is due to the ordered arrangement of monomer and comonomer units in the long chain configuration of the polymer molecular structure. The high crystallinity of the polypropylene results from close packing of the monomer segments. The desired form of the homopolypropylene is the isotactic arrangement. In general, the level of the crystallinity of a pure isotactic polymer is about 93 percent. However, the commercial material contains only 60 to 70 percent crystallinity. This is due to the presence of large amounts of atactic polypropylene in the production scale grades. The atactic polypropylene decreases the ability of the isotactic polypropylene to achieve maximum crystallinity. Furthermore, a reduction in physical properties results from the lower crystallinity of the isotactic polypropylene.

The main properties of polypropylene as compared with other polymers are:

1) low density (0.9 g/ml)
2) high hardness, abrasion resistance and rigidity
3) good heat resistance up to 150°C if the polymer is not subjected to mechanical stresses
4) good impact strength even at low temperature (copolymer only)
5) excellent processibility
6) excellent to good chemical resistance
7) excellent moisture resistance

However, polypropylene is sensitive to thermal and oxidative degradation due to the large number of molecular branch points. In the polymeric chain, the abstraction of a hydrogen atom from a tertiary carbon by a radical is energetically favorable. The order of removal of a hydrogen from the polymer chain is:

$$-\overset{|}{\underset{|}{C}}-H \;>\; -\overset{CH_3}{\overset{|}{CH_2}} \;>\; CH_3-$$

The fragments resulting from the degradation of the polypropylene chain show little tendency to combine with other chains to form long branches. In other words, polypropylene chains can be broken down into small molecules at high temperatures and even at normal processing temperatures. Because of this sensitivity, polypropylene must be well stabilized for optimum processibility and performance. In may cases, an additional stabilizer is required in order to minimize long-term degradation.

Other properties including specific heat, heat of fusion, thermal conductivity, thermal coefficient of expansion, and optical properties are also of practical interest. All these properties are determined by the molecular structure of the polymer.

METHODS FOR THE CHARACTERIZATION OF POLYPROPYLENE

Polypropylene can be characterized through either fundamental or physical properties. The former indicates the molecular structure of the polymer, whereas the latter is related to the performance of the polymer. In general, the characterization of polypropylene can be classified as follows:

1) Molecular Weight

The molecular weight of polypropylene is generally expressed by $\overline{M}_n$ (number average), $\overline{M}_w$ (weight average), and $[\eta]$ (intrinsic viscosity). The DSV (dilute solution viscosity) is occasionally used as an expression of the molecular weight of the polymer. In commercial polymers, the melt flow index is used. A polymer with a high flow rate (low molecular weight) is easily processed. $\overline{M}_n$ is usually determined by the GPC technique. $\overline{M}_w$ is determined by light scattering and GPC while $[\eta]$ and DSV are determined by a viscometer. Information regarding the technique and calculation for polypropylene has been described by Billmeyer[11].

2) Molecular Weight Distribution

The molecular weight distribution can be determined by (a) solubility fractionation, or (b) gel permeation chromatography (GPC). The solubility fractionation method requires the precipitation of a polymer onto a column support and then elution by a gradient solution (mixture of increasingly higher solvency). The GPC method requires eluting a polymer solution through a porous bed. A differential curve of molecular size by a continuous monitoring of the change in refractive index of the eluting solvent is obtained. The $\overline{M}_w/\overline{M}_n$ is used for the expression of the molecular weight distribution of the polymer. It is believed that the GPC technique would serve better for the characterization of polypropylene (easy and rapid technique). This characterization is important since the molecular weight distribution can have a significant effect on the processibility and end-use properties of the polymer.

3) Microstructure

The relative amounts of isotactic, syndiotactic, or atactic structure can be determined by high-resolution NMR. A review paper regarding the determination of polypropylene by NMR has been published recently by Tonelli and Schilling[12].

4) Thermal Analysis

Differential thermal analysis (DTA) has been used effectively in the determination of glass transition temperatures (Tg) and melting temperatures (Tm) of the polymer. It is believed that such an instrument will continue to be useful in the characterization of polypropylene.

5) Crystallinity

The determination of the crystallinity of a polymer is affected by the measuring method. Fortunately, independent techniques lead to unexpectedly good experimental agreement in many polymers. Currently, there are four methods of determining crystallinity: specific volume, x-ray diffraction, heat content as a formation of temperature (DSC), and infrared spectroscopy. It is believed that both DSC and x-ray diffraction methods are attractive due to the advantages of speed and low cost.

6) Melt Flow Index (MFI)

Flow properties for polypropylene resins are measured by a standard test defined by (ASTM D 1238). The term Melt Flow Index, or MFI, is used to describe this procedure for polypropylene that flows through a standard size orifice under a given weight and at a standard temperature. Resins which have relatively short average molecular chain lengths flow through the orifice more easily and are said to have a "high MFI." However, longer chain molecules are more resistant to flow and have a "low MFI."

The Melt Flow Index provides a rough indication of the average molecular weight and processibility of a particular material. Viscous materials which exhibit low MFI values (<2) are used primarily in extrusion processes. Materials with Melt Flow Indices of 2-8 are used in film and many fiber applications. MFI of 8-35 or higher are used for extrusion coating and injection molding of thin wall parts. This information can usually be obtained from suppliers.

Other analytical information, such as comonomer ratio (if any), the rate of crystallization, etc., is also useful. However, this paper will not discuss these subjects. It is important to note, however, that nucleating agents are added to polypropylene in order to increase the rate of crystallization.

PROCESSING CONDITIONS

Polypropylene can be processed by any of the methods ordinarily used for thermoplastics. Slight adjustments in operating conditions are necessary when switching from one grade of polymer to another since each grade of the resin has its own optimum operating conditions.

Polypropylene used for injection molding is operated at a cylinder temperature of 230 to 300°C and a mold temperature below 60°C. Injection pressures should be kept to the minimum necessary for complete mold filling, so that stresses in the moldings are as low as possible.

The polymer for injection molding prefers to have Melt Flow Indices of 4-8. However, the Melt Flow Index of 15 is used for moldings where particularly long flow paths and thin sections are involved.

The choice of polypropylene used for extrusion varies depending on the type of equipment, the shape and the dimensions of the extrudate, and the rheological characteristics of the polymer. In general, the barrel temperatures for most extruded objects, other than sheet, film, and filaments, are 200 to 300°C. For sheet, the barrel temperatures are 220 to 250°C.

The Melt Flow Index for extrusion is normally less than 2.

More detailed information regarding processing conditions is published by the manufacturers.

APPLICATIONS

Although processing knowledge of polypropylene is still in the infancy stage, in some areas of thermoforming, oriented blow molding, and direct injecting foaming, extrusion processing is well defined. Thus, polypropylene is ordinarily extrusion processed. Polypropylene has now been used in areas of automotive parts, plastic pipe, pipe liners and coatings, toys, packaging, household appliances, parts for electronics and electrical industries, and many other fillers. It is expected that the applications of polypropylene will continue to grow due to the good balance of properties, versatility in response to modifications and favorable price/density ratios.

POLYPROPYLENE PRODUCERS

Polypropylene has been produced by several dozens of companies throughout the world. As of 1980, the polypropylene production in the U.S. as well as in other parts of the world, including the name of the company and estimation of capacity (if listed), is shown as follows:

I. U.S. Producer	Plant Capacities (Million lbs/year)
Amoco	830
ARCO	400
Dart (including Rexene)	300
Diamond Shamrock	<200
Exxon	550
Gulf Oil	400
Hercules	1,150
Northern Petrochemical	200
Novemont (USS)	430
Phillips	200
Shell Chemical	600
Soltex	200
Texas Eastman	140
USI	?

II. Foreign Producers

ICI (England)
Solvay
Montedison (Italy)
Sumitomo (Japan)
Hul AG (West Germany)
BASF
Mitsui Petrochemical (Japan)

REFERENCES

[1] J. L. Williams, T. S. Dunn, H. Sugg and V. T. Stannett, Adv. Chem. Ser. 169, 142 (1978).

[2] A. A. Benderly and B. S. Bernstein, J. Appl. Polym. Sci., 13, 505 (1969).

[3] W. K. Busfield, European Polymer J., 17, 333 (1981).

[4] J. Boor, Jr. and E. A. Youngman, J. Poly. Science, Part A-1, 4, 1861 (1966).

[5] E. J. Vandenberg and B. C. Repka, in High Polymers, Vol. XXIX, C. E. Schildkneckt and I. Skeist, Eds., Wiley-Interscience, New York, 1977, p. 337.

[6] Hercules Mixer, 41, 1-10 (Jan. 1958).

[7] A. A. Harban, J. A. Delap and C. W. Moberly, Br. Pat. 940,178 (1963).

[8] C. G. Houser, J. Mitacek, R. G. Rohlfing and W. E. Payne, U.S. Pat. 3,203,943 (1965).

[9] W. M. Saltman, The Stereo Rubbers, John Wiley & Sons, New York (1977).

[10] J. Boor, Jr.

[11] F. W. Billmeyer, Jr., Textbook of Polymer Science, John Wiley & Sons, Inc., New York, 1971, Chapter 3, pp. 62-104.

[12] A. E. Tonelli and F. C Schilling, Acc. Chem. Res., 14, 233 (1981).

R. H. Marchessault

HISTORY OF HELICAL STRUCTURES IN VINYL POLYMERS

1. INTRODUCTION

The key to the understanding of polymer properties is the principle of rotational isomeric states. Each state is weighted according to an internal energy which depends on intramolecular non-bonded interactions. Each chemically regular, linear polymer species has a minimum energy state wherein the torsional angles defining this rotational isomeric state are repeated regularly along the polymer chain. This leads to a helical conformation which is generally the same as found in the crystalline polymer. X-ray diffraction has been the single most useful experimental tool for providing data concerning this "crystalline" minimum energy state. However, the interpretation of the data derived from the x-ray fiber diagram of a crystalline vinyl polymer has evolved gradually as concepts:

- o of polymer chemistry and stereochemistry,
- o of molecular geometry and conformation,
- o of helical macromolecular structure

became entrenched in the thinking of chemists, physicists and biologists. The following is meant to trace some key events in the evolution of concepts relating to the appreciation and acceptance of helical conformations of vinyl polymers. The culmination is found in the brilliant studies of the Natta school(1) on stereoregular vinyl polymers, however the key ideas were clearly discernible long before.

2. STEREOCHEMISTRY OF VINYL CHAINS

In his famous treatise: Die Hochmolekularen Organische Verbindungen, Staudinger(2) anticipated the configurational aspects of today's stereoregular polymers. Discussing the crystallinity of linear macromolecules, he states that they

R. B. Seymour and T. Cheng (eds.), History of Polyolefins, 131–142.

do crystallize if their structure is sufficiently regular. He goes on to state that the possibility of diastereoisomers in polystyrene makes them non-regular which he illustrates with a formula (Fig. 1). Today this formula would be a demonstration of "atacticity". The same ideas also were discussed by Staudinger as early as 1929.

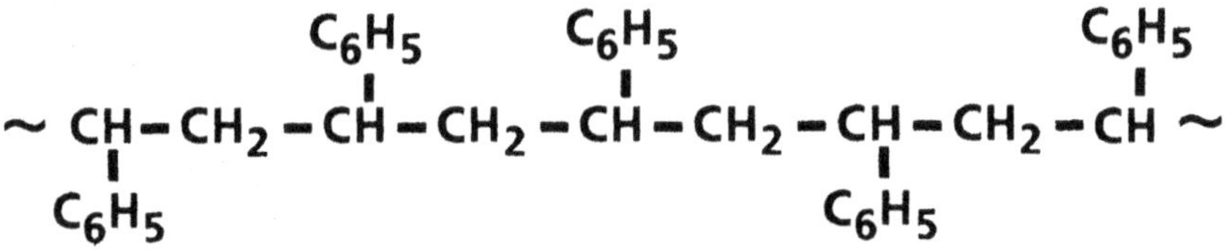

Figure 1. Illustration of diastereoisomerism in polystyrene according to Staudinger(2).

Huggins(3) and Marvel(4) also called attention to asymmetric carbon atoms in vinyl polymer chains. The former observed that as a function of lowering the polymerization temperature in a free radical system one might expect:

> "some regular sequence of dispositions of H and R (where a vinyl chain is ~CH_2-CHR~) such as one in which all of the R groups would be on the same side of the zigzag carbon chain if the molecule were stretched out or one in which the R groups alternated from one side to the other."

The anticipation of "isotactic" and "syndiotactic" is rather clear. Huggins goes on to mention that randomness in H & R is probably responsible (in part) for the poor degree of crystallinity shown by vinyl polymers up to that time. Flory (5) makes a clear statement on the subject in the early part of his 1953 book. The statement is didactic and without reference, implying that this was current knowledge in the early 1950's.

Thus, the stage was set for the discovery of Schildknecht and co-workers(5,6) that cationically polymerized polyvinyl alkyl ethers were crystalline and yielded fiber diagrams. In the lengthy paper published in Industrial and Engineering Chemistry in 1948, the concept of "spatial isomerism" is invoked and the observed 6.2Å fiber repeat is taken to mean that the crystalline chain segments are not fully extended. Their shape was thought to be determined by the sequence of d and l alternation in the asymmetric carbon atoms.

In a footnote statement by Flory(7) in his book, Principles of Polymer Chemistry, he comments on Schildknecht's hypothesis in terms:

> "The opposite might be true; i.e., sequences of units of like configuration might be preferred".

Thus, the idea of "syndiotactic" and "isotactic" were on the same page of this classic monograph but, without any nomenclature elegance. At this time (the early 1950's), the real examples of polymer stereoregularity had not yet been reconciled with the power of crystallographic interpretations. However, the lack of crystallinity in most synthetic vinyl polymers was conceptually related to configurational irregularity at that time. Nowhere is this more obvious than in the papers of the ICI group(8) of which Bunn(9) is the best spokesman for vinyl polymers:

> ". . . the typical noncrystalline polymers are those in which there is either chemical or stereochemical irregularity or both; . . . and among homopolymers it is those in which there are likely to be left- and right-hand groups indiscriminately along the chains which are noncrystalline - for example polystyrene, polymethyl methacrylate . . ."(9)

Thus, before the isotactic polymers became available, the importance of stereochemistry for vinyl polymer crystallization was appreciated. However, the lack of special polymerization conditions which would lead the vinyl polymers to:

> "have their substituents in corresponding or regularly alternating position"(9)

meant that there was not the wealth of examples to make this topic an important issue among polymer chemists. Examples of crystalline vinyl polymers, such as, polyvinyl chloride and polyvinyl alcohol, were explained in terms which were in sympathy with the subsequent development of stereoregular polymer systems.

3. EARLY CRYSTALLOGRAPHY OF VINYL CHAINS

Throughout the 1930's, the basic concepts for interpreting fiber diagrams were settled. Because of a lack of model building methods, using high-speed computers, interpretation of data was often difficult except in cases such as polyethylene and other planar zig-zag situations for vinyl chains. Likewise, a lack of basic crystallographic data on suitable model compounds and on a range of related organic molecules and functional groups resulted in some uncertainty in the interpretation of the crystalline conformation of macromolecules which were not clearly planar zig-zag structures.

By the early 1940's, helical concepts for polymer chain structures were present in the minds of crystallographers who worked with crystalline polymers. Notable among these was the study of polyisobutylene by Fuller, Frosch and Pape(10) which appeared in 1940. Following on the preliminary study by Brill and Halle(11), who had concluded that:

> "the molecules in the stretched state are not planar zig-zag, but because of the methyl side-groups are probably helical",

Fuller et al. examined a variety of helical arrangements for this elastomer which gives a very beautiful fiber diagram under tension. By today's standards, their detailed interpretation must be considered inconclusive, but the helical analysis was sound and it was a pioneering effort in the field of crystalline vinyl polymers. It gave strong support to the concept of head to tail polymerization and shows that interpretation of vinyl structures in terms of helices was prevalent by the late '30's. Fig. 2 shows the sort of elegant diagam which accompanied this paper.(10)

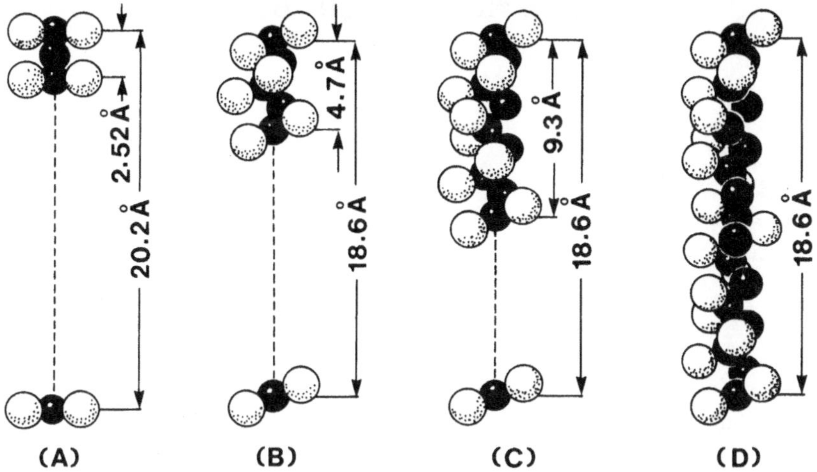

Figure 2. Schematic of polyisobutylene chain from Fuller, Frosch and Pape(10) (the larger spheres represent the methyl groups). The observed fiber repeat of 18.6Å was not an integral of 2.52Å hence a helical arrangement was proposed which includes 8 repeating units per turn. Various 1:3 chains: (A) planar zig-zag; (B) to (D) non-planar forms which produce the required shortening.

This effort was shortly followed by Bunn's (12) masterful paper which elucidated the principle of "staggered bonds". This dealt with the question of the most stable bond position (dihedral angle) in a succession of singly-linked carbon atoms. The paper showed a remarkable grasp of the early concepts of conformational analysis and provided the basic thinking which is still operative today in polymer crystallography. This particular paper predicted the familiar "trans gauche" sequence in crystalline vinyl chains leading to a helix with three residues per turn and a 6.2Å repeat (Fig. 3). This paper was cited by Natta(13) in his lecture on Isotactic Polymers at the Zurich IUPAC Symposium on Macromolecular Chemistry in 1955. Ironically, 6.2Å was also the fiber repeat observed by Schildknecht, et al.,(6) for polyvinyl isobutyl ether, but (as stated above) they favoured an alternation of d, l asymmetric carbons as the crystallizing sequence.

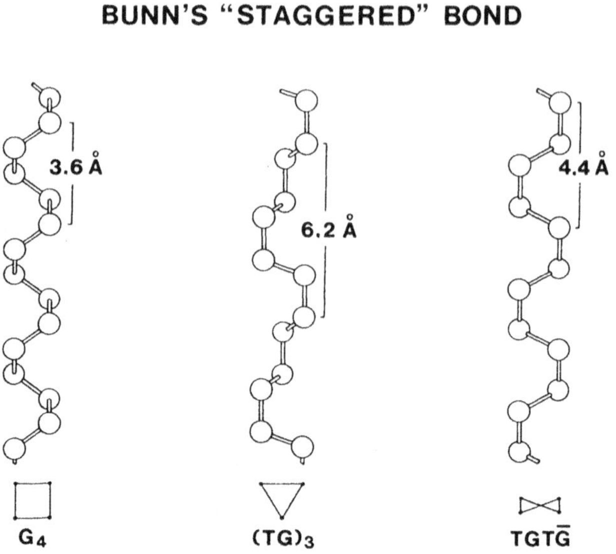

Figure 3. Classic diagram from Bunn(12) which illustrates the various chain conformations which can be derived from various sequences of "trans" and "gauche" arrangements.

The early 1950's also saw helical conformations proposed for two important polymers Teflon and Kel-F.(14) The fiber repeat for the latter was about 35Å, which was interpreted in terms of a slowly turning 14 residue helix, indicating that the helical concept was now well established.

4. THE ISOTACTIC POLYOLEFINS

By 1955, the Natta school had not only proposed the threefold helical structure of polypropylene, but found that poly-α-butene and polystyrene also had the same fiber repeat.(13) Thus the principle of the staggered arrangement of carbon-carbon bonds as originally proposed by Bunn(12) was given strong experimental support. From this broad base of crystalline polymer conformation data along with crystallographic symmetry considerations, the isotactic concepts of poly(α-olefin) structures were defined. Natta(15,16) referred to natural polymers as "precisely constructed models" whereas the early synthetic polymers were considered as an "irregular and disordered association of constitutive elements". Since crystalline natural polymers contain asymmetric carbons, stereochemistry appears as the fundamental influence on polymer crystallization. From there, Natta went on to state the basic rule of polymer helix generation on crystallization: **the different structural units . . . in the identity period occupy equivalent positions about an axis.** Since the arrangement of the structural units along an istoactic polymer helix are generated by translation and rotation, a crystallizing poly(α-olefin) must have the same pseudo-asymmetry along the chain. Without this condition, successive monomers cannot superimpose on each other when translated and rotated. Thus, the concept of crystallizability and steroregularity are irrevocably tied together.

Significantly, in the U.S. Patent Office the importance of stereoregular poly(α-olefins) was appreciated only in terms of crystallinity. The "count" in the interference set up by the U.S. Patent Office in 1954 between patent applications assigned to Montedison S.p.A., Dupont de Nemours and Company, Phillips Petroleum Company and Standard Oil Company (Indiana) reads:

"Normally solid polypropylene, consisting essentially of recurring propylene units, having a substantial crystalline polypropylene content."

This placed emphasis on the necessary condition for crystallinity, i.e., recurring propylene units, but overlooked the ultimate condition viz., stereoregularity.

The stereoregularity statement came shortly after (15) backed by sophisticated evaluation of x-ray data. However, this sort of support was lacking in the patent applications which made up the "interference case". In fact, only two of the applications (Montedison and Standard Oil) could reach back to research studies that included x-ray fiber data and interpretation in terms of the familiar threefold helix predicted by Bunn on the principle of "staggered bonds". From preliminary data in early 1954, the Natta studies went on to provide the classical publications in the field of isotactic polymers. With all due credit to his equally famous collaborators, notably Corradini and Pino, Natta must be regarded as one of the premier inventors in the polymer field. His early brush with a sample of crystalline polystyrene (17) in 1936, apparently an accidentally produced crystalline sample given to him by some polymer chemist, set him off on a lifelong search for crystalline vinyl polymers. When they were prepared in his laboratory, he did not take long to recognize what he had.

By contrast, the other early protagonist of the threefold polypropylene helix was inexperienced with crystalline polymers. He was Wayne Kimball, a young M.Sc. graduate responsible for running the x-ray equipment at Standard Oil (Indiana). He recognized the importance of obtaining a fiber diagram after diffractometer traces of Standard's polypropylene in the 1953 timeframe showed "substantial crystallinity". (Fig. 4). From a cold stretched film made of xylene insoluble polypropylene a typical low order but well-oriented fiber diagram yielded the pattern shown in Fig. 5. The numbered reflections led Kimball to propose an orthorhombic unit cell:

$$a = 12.6\text{Å} \quad b = 10.6\text{Å} \quad c(\text{fiber axis}) = 6.25\text{Å}$$

He concluded that the cell had 4 chains with a total of 12 propylene units, i.e., threefold helical.

Kimball got his inspiration from the polyisobutylene study Fuller, Frosch and Pape (10). Other than that, he seems to have worked in isolation and obvious approaches, such as, hot annealing to improve crystallinity were not undertaken. The writer has undertaken an evaluation of the Kimball data (Fig. 5) and comes to nearly the same results. Use of a Bunn chart for indexing leads also to the same hkl assignments. The fiber repeat of 6.25Å is lower than the 6.4Å report by Natta et al. because reflection no. 2 in Fig. 5 is a pseudo-meridional reflection. Tilting the fiber when the x-ray pattern was recorded would have given a slightly larger value for the fiber repeat.

5. CONCLUSIONS

One of the evident factors in the success of the Natta school in the elucidation of the principles of polymer stereoregularity is that they were first and foremost

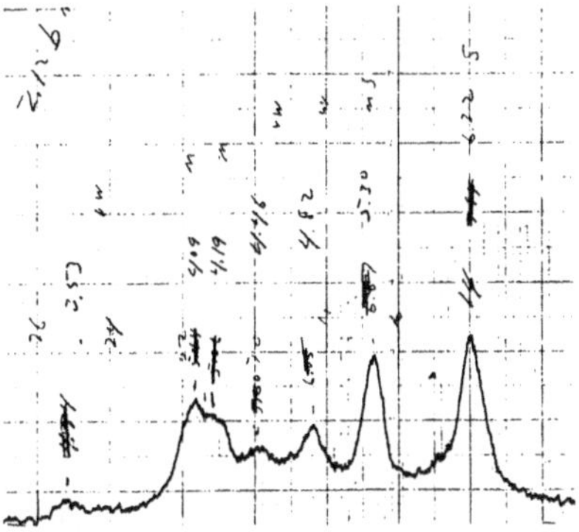

Figure 4. Diffractometer trace crystalline polypropylene from Standardil of Indiana from 1953.

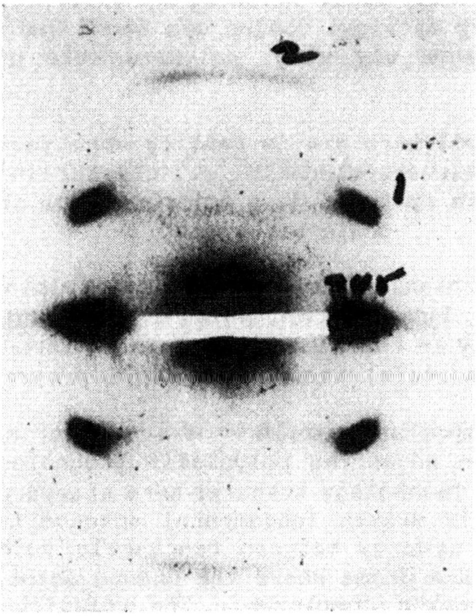

Figure 5. Fiber Diagram of cold drawn crystalline polypropylene from which Wayne Kimball derived a polypropylene unit cell and proposed threefold chain symmetry in March, 1954.

chemists. It was this chemical background which intuitively led them to solvent extract the first synthesized stereoregular polymers (which had all the appearances of amorphous elastomers) in order to remove the non-regular fraction. Unfortunately, polymer chemists by this time had perhaps become used to living with chemical heterogeneity. There was not enough rigid organic chemistry. The standards which were the hallmark of classical organic syntheses and small molecule crystallography in the 1950's and onwards were only starting to be adopted by polymer chemists. The major developments in the two latter fields and the masterpieces of protein structure analyses(18), became the foundation for the proper interpretation of vinyl polymer helices.

In retrospect, the attention of scientists is in the field where rapid progress is possible. Until the advent of Ziegler-Natta catalysis and high resolution NMR, tacticity of vinyl chains was not a major industrial research topic. When leaders in the chemistry of polymers gave some attention to the matter, their basic thinking and interpretation of the stereochemistry of vinyl chains was sound and they clearly taught the reason why vinyl polymers were generally not crystalline:

> "Such polymers are in reality copolymers of <u>d</u> and <u>l</u> structural units ... crystallinity may occur in spite of <u>d</u>, <u>l</u> heterogeneity in the chain structure"(7)

No new thinking was required to explain crystalline vinyl polymers, but the combination of chemistry and polymer crystallography in the laboratory under a strong leader was probably an essential recipe for rapid progress.

The important role of industrial research laboratories in advancing polyolefin technology cannot be overstressed. Technology transfer sets standards of utility which ultimately drives fundamental science to an ordered plateau. The synergy between benchscale science and its applications is a drama where the unsung actors are in the industrial research complexes. The relation between the Natta School at Milan and the Montedison S.p.A. is a classic example of industry-university collaboration in a focussed manner. This type of interaction, i.e., sponsorship of a university school by a corporation, is being imitated in the life sciences by long term agreements, such as, those announced by Hoechst, Dupont and Monsanto.

The importance of scientific excellence in the industrial laboratory is one lesson which should be learned from the polypropylene interference trial. Kimball at Standard Oil of Indiana was a voice in the wilderness. If the oil industry had decided to synthesize polyolefins, their first preoccupation should have been scientific leadership in the field of polymers, perhaps by hiring Carothers-type scholars. Such an approach provided Dupont with a long tradition in polymer development but, it failed them when it came to isotactic polypropylene because there was not a focussed strategy to capture the field.

Xerox Research Centre of Canada
2660 Speakman Drive
Mississauga, Ontario L5K 2L1

REFERENCES

1. G. Natta, J. Polymer Sci., 16, 143 (1955).

2. H. Staudinger, Die Hockmolekularen Organischen Verbindungen, Julius Springer, Berlin, pp. 162-165 (1932); Helv. Chim. Acta, 12, 947 (1929).

3. M. L. Huggins, J. Am. Chem. Soc., 66, 1991 (1944).

4. C. S. Marvel, Chemistry of Large Molecules, ed. by Burke and Grummitt, Interscience Pub., New York, p. 240 (1943).

5. Paul J. Flory, Principles of Polymer Chemistry, Cornell University Press, Ithaca, N. Y., p. 56 (1953).

6. C. E. Schildknecht, S. T. Gross, H. R. Davidson, J. M. Lambert and A. O. Zoss, Ind. Eng. Chem., 40, 2104 (1948).

7. Paul J. Flory, Principles of Polymer Chemistry, Cornell University Press, Ithaca, N. Y., p. 238 (1953).

8. R. Hill and E. E. Walker, J. Pol. Sci. 3, No. 5, 609 (1948).

9. C. W. Bunn, J. Appl. Phys., 25, 820 (1954).

10. C. S. Fuller, C. J. Frosch and N. R. Pape. J. Am. Chem. Soc., 62, 1905 (1940).

11. Brill and Halle, Naturwischenshaften, 26, 12 (1938).

12. C. W. Bunn, Proc. Royal Soc., (London) A180, 67 (1942).

13. G. Natta, J. Am. Chem. Soc., 77, 1708 (1955).

14. H. S. Kaufman, J. Am. Chem. Soc., 75, 1477 (1953).

15. G. Natta and P. Corradini Supp. Al Volume XV, Series X, Del Nuovo Cimento, 9 (1960).

16. G. Natta, Scientific American, 205, No. 2, 33 (1961).

17. Giulio Natta: Present Significance of his Scientific Contribution, ed. by Sergio Carra, Federico Parisi, Italo Pasquon, Piero Pino, Editrice di Chimica s.r.l., p. 11 (1982).

18. W. L. Bragg, J. C. Kendrew and M. F. Perutz, Proc. Roy. Soc. (London), A203, 321 (1950).

Erik G. M. Tornqvist

POLYOLEFIN ELASTOMERS - FIFTY YEARS OF PROGRESS

INTRODUCTION

Most commercially valuable synthetic elastomers are hydrocarbons made form unsaturated monomers. All such monomers may be considered olefins and, if so, the scope of this presentation would be exceedingly large. However, for the purpose of this paper we shall use a more restrictive definition and limit ourselves to elastomers made primarily from monoolefins, either by direct synthesis or by postpolymerization modification of a polyolefin which in itself may or may not be elastomeric. Such a restriction brings the contents of this paper more in line with those of the other ones in this symposium, but it does, of course, leave out the most important group of hydrocarbon elastomers, the polydienes, as shown in Figure 1.

Figure 1

SCOPE OF PRESENTATION

Included

- Elastomeric materials from direct polymerization of olefins: Polyisobutylene, Butyl Rubber (IIR), and Ethylene-Propylene Elastomers (EPM and EPDM).

- Elastomers from postpolymerization modification of polyolefins: Bromo- and Chlorobutyl (BIIR and CIIR), and Chlorosulfonated Polyethylene (CSM).

Not Included

- Elastomers from dienes as major component, e.g., Polyisoprene (IR), Polybutadiene (BR), and Styrene-Butadiene Rubber (SBR).

- Elastomers from chemically modified olefins, e.g., Fluorocarbon Elastomers and Polychloroprene (CR).

R. B. Seymour and T. Cheng (eds.), History of Polyolefins, 143–161.

POLYISOBUTYLENE, BUTYL RUBBER AND HALOBUTYL

In a sense, the beginning of polyolefin elastomers (as defined) can be traced back to work published in 1873 by Butlerov and Goryainov(1), who found that isobutylene polymerized to liquid products under the influence of boron trifluoride. However, this is only a post factum relationship. The investigations that eventually led to the first practically useful polyolefin elastomer were started in Germany in the late 1920's by Otto(2). Supposedly, they were inspired by work carried out many years earlier by Hofmann, one of the foremost pioneers of synthetic elastomers(3). However, Otto's studies were not directed toward making this kind of product. Rather he was interested more generally in studying polymers that could be obtained from the action of boron trifluoride on olefins. He did look at isobutylene briefly and noted its high reactivity, but ethylene was the only monomer studied in greater detail. By this time the first of Staudinger's many epoch-making papers on polymerization reactions and polymer synthesis had appeared(4-6), and Otto recognized that, in the absence of side reactions, ethylene should produce a mixture of alpha-olefins of widely varying molecular size but belonging to the same (even-carbon-numbered) homologous series. The polymerization product could therefore be expected to have petroleum-like properties and, if in the right viscosity range, to be similar to high quality lubricating oils. He also observed that the molecular weight of the polymer (as measured by viscosity) was strongly influenced by the reaction temperature and that it increased strongly with decreasing temperature, especially below about 15°C.

Apparently, I. G. Farben, the large German chemical combine, became interested in Otto's work and he moved from the Coal Research Institute at Breslau to the I. G. facilities at Ludvigshafen-on-the-Rhine (now the main site of B.A.S.F.). Judging from the filing date of the patents that followed, this must have happened around 1929 or 1930. Together with Mueller-Cunradi, he began to look more closely at isobutylene. They soon determined that this monomer also produced polymers of increasingly high mol. wt. as the temperature of polymerization was lowered. By using temperatures as low as -80°C, they were able to obtain polyisobutylene having an average mol. wt. greater than

5,000[7][1]. They also found that the clear sticky, gum-like, slow-flowing material thus obtained exhibited interesting properties both as a thickener and as a viscosity index improver for lubricating oils.

At this time something else had happened which was to have a profound influence on the development of the first elastomeric polyolefin. The Standard Oil Co. of New Jersey (then popularly called "Jersey" and now renamed Exxon Corporation) had become interested in the research advances made by I. G. Farben in a wide variety of areas important to an oil company. Contacts between the two companies were established in the late 1920's, primarily on the initiative of F. A. Howard[9], President of the Standard Oil Development Co. A formal agreement according to which the companies were to cooperate in the development of new products and processes based on raw materials from petroleum was signed in 1930. As a consequence, the advances in the polymerization of isobutylene were rapidly brought to the attention of Jersey management, and arrangements were made for Otto to come to the Bayway Refinery in Linden, N. J., and help in the commercial development of polyisobutylene-based oil additives. He arrived in the United States around 1933 and stayed for about two years according to Thomas[10], who became one of Otto's assistants. The great progress made during this time period has been described in considerable detail by Thomas in his Goodyear Medal Address of 1969[10], and it may suffice to list just a few of the advances. A pilot plant was built, which made possible a systematic study of reaction variables, improved methods of purifying isobutylene were developed and liquid ethylene was found to be the most suitable reaction medium. These accomplishments may not seem spectacular by present day standards, but they were made under very difficult conditions. Most of the work was carried out on the roof of a three-story laboratory building where at least good ventilation, required by the use of liquid ethylene and boron triflouride, did not present too much of a problem.

---

1 According to the original Staudinger scale for $\overline{M}_v$ but probably in reality about ten times higher.

At the time of Otto's return to Germany, the standard polymerization procedure was to add a solution of boron triflouride in liquid ethylene to a mixture of isobutylene and ethylene in the reactor. Bannon[11] working with Thomas reversed this procedure by adding the ethylene-isobutylene mixture to a solution of a small amount of boron triflouride in ethylene and thus established the concept of cold slurry operation, which is still in use today for making both polyisobutylene and butyl rubber. Interestingly, Otto and coworkers[12] in Germany developed an entirely different technique involving an endless belt for commercial production of polyisobutylene.

Although exhibiting certain elastomeric properties, this material, which is still being produced in both the U.S. (under the name of Vistanex®) and in Germany (as Oppanol®), is not an elastomer in the fullest sense, since it cannot be cured. However, it does find use in a number of applications in competition with true elastomers, such as in pressure sensitive adhesives, polyblends and non-cured sheeting. Together with polystyrene it was also for a long time a favorite material for studying various aspects of high polymer physics.

A patent disclosing the use of complexes of $AlCl_3$ with polar solvents for polymerization of isobutylene[13] came to Thomas' attention shortly before Otto left for Germany. However, the complexes disclosed produced only low mol.wt. polyisobutylene, but they did cause Thomas to look more closely at $AlCl_3$ as a polymerization catalyst. His observations led him to conclude that this compound was capable of producing high mol. wt. polyisobutylene, but that the insoluble particles became coated with polymer, which then prevented further reaction. This problem was solved when one of his colleagues, Slotterbeck, mentioned that $AlCl_3$ was surprisingly soluble in ethyl chloride. They rapidly established that $AlCl_3$ in alkyl halide solution is a very effective catalyst for low temperature polymerizations[14], and eventually $AlCl_3$ dissolved in methyl chloride became the standard catalyst for isobutylene polymerization.

This was the situation the last weekend in July 1937, when Thomas and Sparks had come to their laboratories to discuss and try some new ideas. They decided to try to overcome the main shortcoming of polyisobutylene through the incorporation of a diolefin, which would provide some unsaturation and thus make the polymer vulcanizable. Their first experiments, which also involved some improvised vulcanization, did not produce particularly impressive looking products, but they did indicate that copolymerization had taken place. Improvements in the polymerization procedure followed rapidly and butyl rubber was born[15]. Its inventors are shown in Figure 2. However, numerous problems had to be solved before any commercialization could even be suggested[16]. It is not possible to account for all of these, but a few may be mentioned.

Figure 2

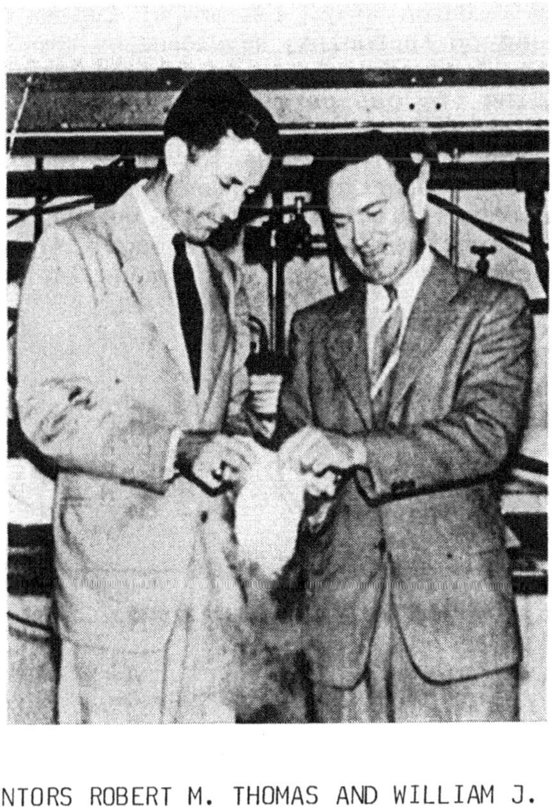

INVENTORS ROBERT M. THOMAS AND WILLIAM J. SPARKS
WITH FIRST BATCH OF BUTYL RUBBER

Because of the low level of unsaturation available as compared to natural rubber (Figure 3) special conditions and recipes for vulcanization had to be developed[17], a problem that has been described in considerable detail by Sparks in his Goodyear Medal Address of 1963(18). It was also discovered that isoprene is a better comonomer than butadiene. Thus a method for isolating this monomer from refinery streams had to be developed. Several problems relating to the polymerization process also had to be overcome, especially with respect to catalyst addition(18,19). However, even with the great improvements made both in the basic polymerization procedure and in the polymer itself, butyl rubber may not have been commercialized for many years had it not been for World War II. The first two plants were built by the governments of the United States at Baton Rouge, LA, and of Canada at Sarnia, Ontario, based on technology developed by the Standard Oil Development Co. Commercial scale production started in 1943. The outstanding low gas permeability exhibited by butyl rubber made it a preferred material for inner tubes and this application was a dominant one for many years, although other outstanding properties, such as aging resistance and good damping characteristics eventually opened up many other applications. Nevertheless, the introduction of tubeless tires threatened to decrease the demand for butyl rubber drastically.

Figure 3

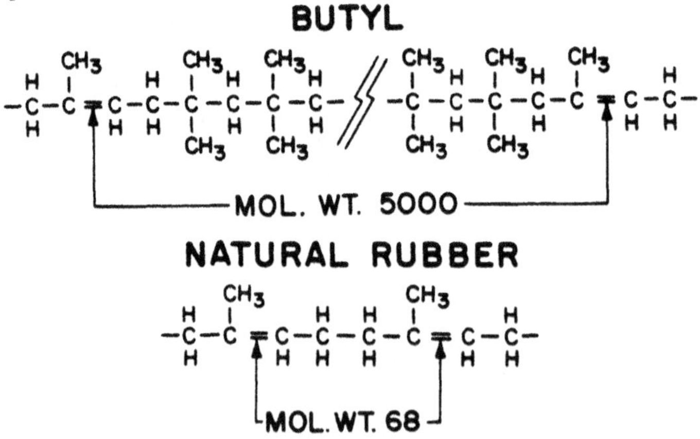

COMPARISON OF STRUCTURES OF BUTYL AND NATURAL RUBBER

Obviously, the good air retention qualities of butyl would still be desirable in a tubeless tire, but the lack of compatibility of this low unsaturation rubber with the more conventional diene-based rubbers normally used for making tires presented a great problem. Researchers at many companies worked on finding a solution and Morrissey[21-26] at the Goodrich Co. finally found a practical one.

He established first that the compatibility problem could be overcome through partial bromination of the rubber[21-24] and then that chlorination would also yield reasonably satisfactory results[25-26]. Small scale manufacturing and marketing of brominated butyl was started in 1954 under the name of Hycar® 2202.

Meanwhile at what is now Exxon, work on modifying butyl had also been progressing. Attempts at chlorinating this polymer had been made, but primarily for the purpose of introducing some polarity. Severe breakdown of the polymer molecules had been noticed at the rather high chlorine levels tried. Baldwin and Thomas[27] decided to employ a more gradual chlorination of butyl rubber in solution. They noticed that initially chlorine was consumed very rapidly and that hydrogen chloride was evolved. Eventually, they established that for each mole of chlorine added, about one mole of HCl was formed[28]. Thus it was indicated that the chlorination resulted in a substitution reaction rather than a simple addition of halogen to double bonds as claimed by Morrissey[25]. If so, the reaction could be expected to produce isolated allylic chlorine rather than a structure involving a pair of chlorines added to adjacent carbon atoms as shown in Figure 4. Since the conditions of chlorination used by Baldwin and Thomas and by Morrissey were different, especially with respect to temperature, different structures were plausible. However, it seems that the difference was more apparent than real. Many of the modern tools for determining the structure of organic compounds were not available to Morrissey in the late 1940's and he apparently relied completely upon iodine number determinations, which indicated little residual unsaturation. As we now know, this was not because of the absence of C=C bonds but because allylic chlorine will make such bonds rather inert toward halogen addition[29]. Today it is well recognized that the major portion of the halogen in both chloro- and bromobutyl is present as Structure II in Figure 4[30-32].

Figure 4

CHLOROBUTYL STRUCTURES

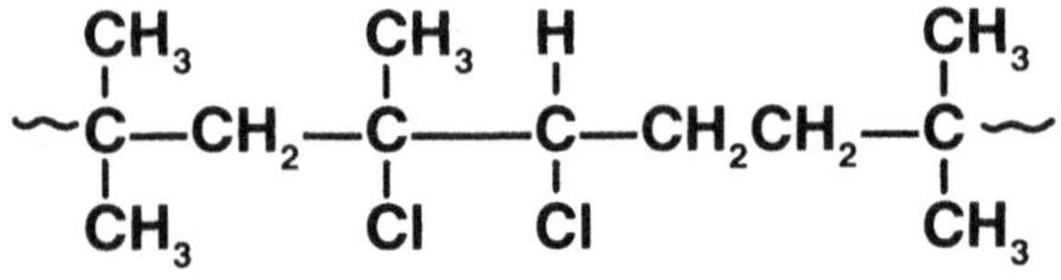

I. Claimed by Morrissey [25]

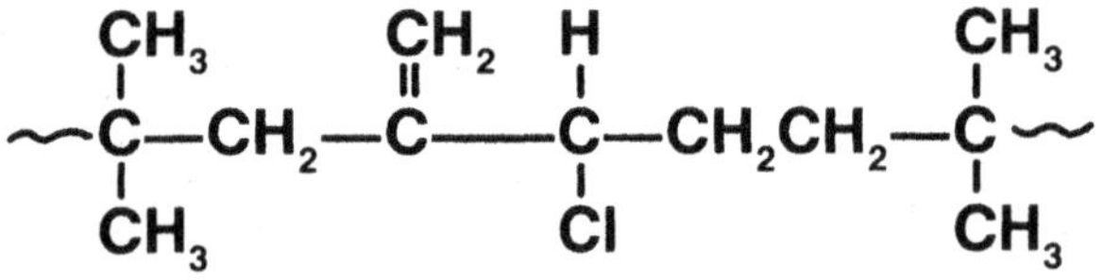

II. Major Component in Commercial Product[30-32]

Subsequently Baldwin and Thomas developed an efficient process for large scale manufacture of chlorobutyl, which was commercialized in 1960. Large scale manufacture of bromobutyl was then started in 1970 by Polysar in Canada.

The successful introduction of halobutyl obviously increased the demand for butyl rubber considerably but meanwhile many new applications for the basic rubber were also found. Therefore, instead of a declining consumption, a rather steady and sharp increase has taken place since its original commercialization as shown by the data for the U.S. consumption 1945-1982 reported in Figure 5.

Figure 5

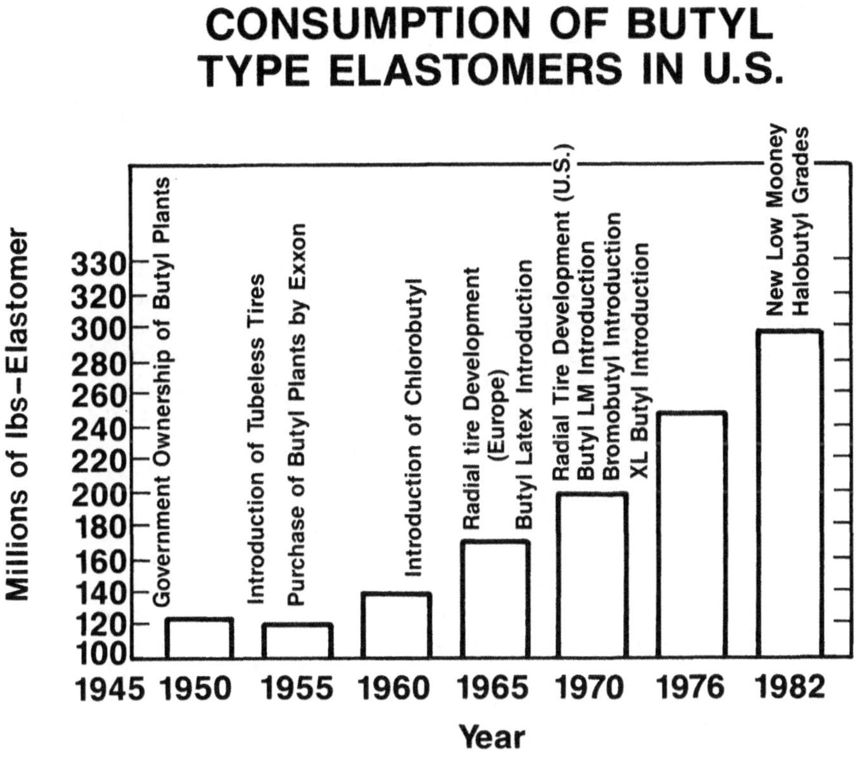

CHLOROSULFONATED POLYETHYLENE

Both compositionally and chronologically, chlorosulfonated polyethylene now fits into the picture. Like halobutyl, it is made by postpolymerization modification of a polyolefin, but in this case the starting material is primarily plastic rather than elastomeric.

It started in the late 1930's when McQueen[33] treated a suspension of low density polyethylene, then a newly discovered material, in carbon tetrachloride with sulfur dioxide and chlorine. This resulted in the introduction of both chlorine and sulfonyl chloride groups into the polymer. He recognized that the product thus obtained might have some direct utility, especially in blends, but his primary objective was to use it as an intermediate for making water-soluble polymers containing sulfonic acid groups. This was accomplished through hydrolysis of the sulfonyl chloride groups with suitable bases.

Some years later, McAlevy and coworkers[34,35] realized that sulfochlorinated polyethylene of the right composition exhibited outstanding properties in a number of applications, especially because of good heat resistance and resistance toward aging in general. They recognized that the chlorine substituents were particularly important to the elastomeric properties, while the sulfonyl chloride groups made the polymers curable with metal oxides. For the polyethylenes available at the time (1947), the optimum elastomeric properties were found to occur at chlorine contents between 27 and 30%[35,36]. McAlevy also found that the best product resulted from sulfochlorination in solution and that peroxides and other free radical producing agents could be used in place of radiation to increase the efficiency of the reaction.

Based on these developments, the DuPont Company started small scale production of chlorosulfonated polyethylene under the name Hypalon® at Belle, WV in 1954. A full scale commercial plant was then built at Beaumont, TX in 1957[37]. By that time, new types of polyethylene had become available, especially through polymerization with what is now known as Ziegler catalysts[38].

Heuse[39] recognized that the higher density, more linear polyethylene now obtainable might produce elastomeric materials at higher chlorine levels than could be used with the earlier polyethylene grades. He demonstrated that chlorine levels in the range of 38% to 48% could be advantageously used. When cured, the polymers thus obtained exhibited outstanding resistance toward swelling by oils and organic solvents. Today, most grades of Hypalon® are based on the latter type of polyethylene. DuPont is still the dominant producer, but some Japanese competition is apparently appearing on the horizon.

ETHYLENE-PROPYLENE BASED ELASTOMERS (EPM AND EPDM)

This brings us to the final type of polyolefin elastomers, namely those based on ethylene and propylene. They came as a direct consequence of Ziegler's catalyst discoveries[38], the history of which is discussed in many other papers at this symposium. Very early in 1954, Natta, as a consultant of the Montecatini Company was informed by Ziegler[40] about the discovery of the new transition metal containing catalysts capable of producing solid polyethylene. The information was disclosed under an agreement between Ziegler and Montecatini that had existed for some years. At that time, Natta[41] had already become so interested in the ethylene addition to pure aluminum alkyls that he had begun studying the kinetics of this reaction. He was in an excellent position, therefore, to take up the study of the new catalysts disclosed by Ziegler, which he apparently did without delay. However, his important discoveries in the field of stereospecific polymerization, both with alpha-olefins (described in other papers at this symposium) and with conjugated dienes[42] caused him at first to concentrate on these subjects. Therefore, work toward making truly elastomeric ethylene-propylene copolymers was not started by Natta and coworkers[43] until some time in 1955. By this time many other companies had learned about the discoveries at Mülheim and had begun to work on various aspects of Ziegler[2] catalysis. As one company publication[45] put it, "The relevations of the Mülheim (Ziegler) catalysts had the same effect as the starting gun of a race in which the laboratories of the interested industries had entered".

By that time, knowledge of structure-property relationships had advanced so much and become so widespread that numerous polymer chemists realized that properly prepared ethylene-propylene copolymers would have interesting elastomeric properties. Actually, McAlevy et al[34] as early as 1946 stated that methyl groups would have a similar effect as chlorine atoms in imparting elastomeric properties to polyethylene. At that time, there was no method of making such a polymer, but copolymerization of propylene with ethylene offered to make it a reality. The low cost of the two monomers did, of course, provide a particular incentive for research in this area.

It appears that Ziegler and coworkers were the first ones to establish that their newly discovered catalysts could copolymerize the two monomers, which happened early in 1954[46]. However, the first publications disclosing attempts at making elastomeric ethylene-propylene copolymers did come from Natta and coworkers[47-52]. They found that

2 Ziegler[44] preferred to call his catalysts Mülheim catalysts for the place where they were discovered.

catalysts based on vanadium[49-51] gave better copolymer properties than those based on titanium[52], and that the best catalysts resulted from the use of a soluble vanadium compound, VCl_4[50] in particular. The latter compound provided catalysts exhibiting the lowest reactivity ratio product, $r_1 \cdot r_2$ 1 (as conventionally defined), resulting in copolymers of the most alternating structure at any given comonomer ratio.

The simultaneous work going on at other laboratories was revealed subsequently, primarily in the patent literature, which is so voluminous that it cannot be treated fairly in a presentation of this scope. Most of the patents have now expired, but it is still a sensitive area. It should be noted, though, that while numerous important improvements have been made over the years and continue to be made, they are all dwarfed in importance by Ziegler's original discoveries, which made it all possible. References to some of the more important early patents can be found in review articles[46,53].

Ethylene-propylene (binary) copolymers were commercialized in 1960 by Exxon. As saturated hydrocarbons, they cannot be cured by conventional sulfur vulcanization. However, unlike polyisobutylene, they are susceptible to peroxide initiated crosslinking without attendant severe chain scission. Therefore, they find use both in unvulcanized and vulcanized state.

Nevertheless, the desirability of producing a sulfur vulcanizable copolymer led to a search for suitable dienes that could be copolymerized with ethylene and propylene, analogous with the copolymerization of isobutylene and isoprene in butyl rubber. Undoubtedly, in many laboratories attempts were first made to use one of the more readily available conjugated dienes, e.g. butadiene, isoprene and piperylene, for this purpose. These were not met with success.

Several suitable diene termonomers were eventually found as shown in Figure 6. Because of its alpha-olefinic unsaturation, 1,4-hexadiene copolymerizes in the normal fashion expected for a Ziegler catalyst. The first EPDM was commercialized with this termonomer by DuPont in 1963.The other three dienes polymerize through the same ring olefin functionality. This may seem surprising in view of the very low reactivity (when observed) of Ziegler catalysts toward internal double bonds. However, the strain from the cyclic (<u>cis</u>) configuration apparently makes the reactivity of this

Figure 6

TERMONOMERS FOR MAKING EPDM AND RESULTING STRUCTURAL UNITS

MONOMER	FUNCTIONALITY
$CH_2 = CHCH_2CH = CHCH_3$ 1,4-Hexadiene (1,4-HD)	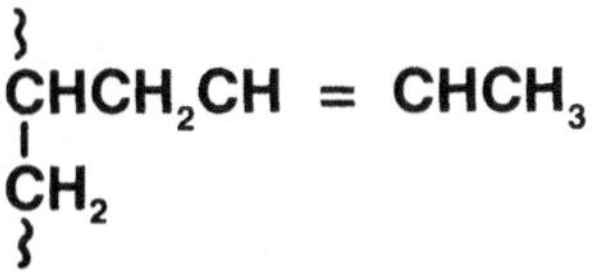
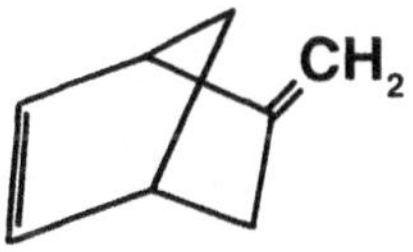 5-Methylene-2-norbornene (MNB)	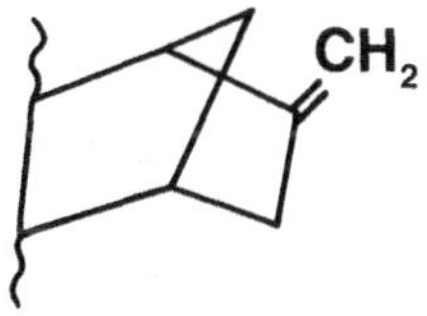
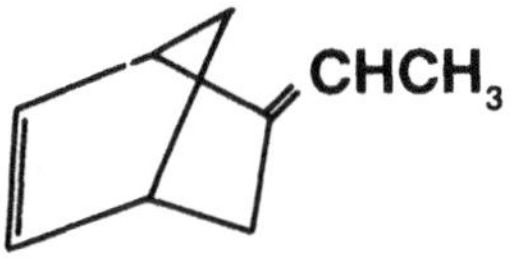 5-Ethylidene-2-norbornene (ENB)	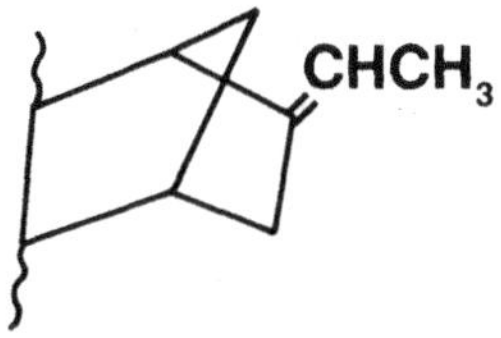
 Dicyclopentadiene (DCP)	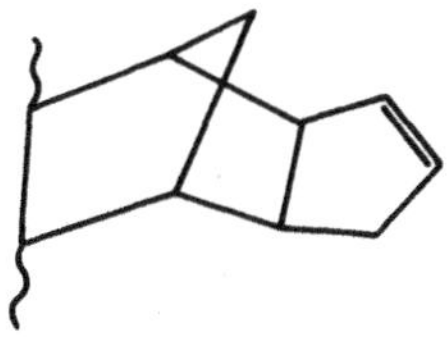

double bond very much greater than would have been the case in a similarly unsaturated linear hydrocarbon. Interestingly, the high reactivity of the double bond in norbornene, was discovered very early by Anderson and coworkers[55,56] at DuPont, long before any of its doubly unsaturated homologs became of interest as termonomers in the manufacture of EPDM[57].

Today EPM and EPDM are produced by at least eleven companies in many more plants around the world. Their superior aging properties stemming from a completely saturated backbone are particularly valued. Some of the important applications are: sheeting, automotive uses such as sponge and hoses, and plastics blending. The rapid increase in production from the beginning in 1960 to 1984 is shown in Figure 7.

Figure 7

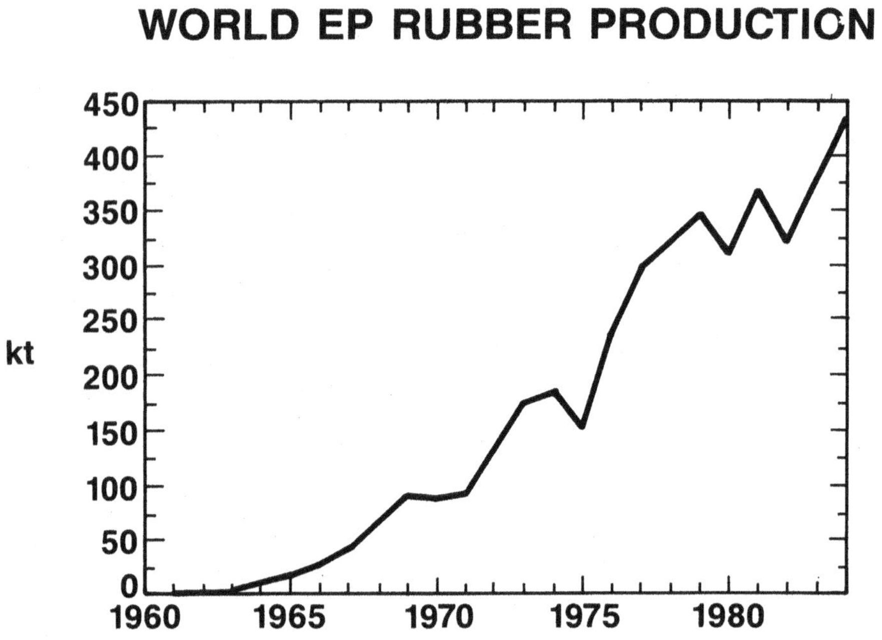

Sources

International Institute of Synthetic Rubber Producers
Exxon Estimate (Prior to 1970)

As the total annual worldwide production of isobutylene-based polymers exceeds 600,000 tons, considerably more than one million tons of polyolefin elastomers are now being produced per year. When one considers that not even high molecular weight polyisobutylene was commercially available 50 years ago, the progress made in this field during the intervening period cannot but be considered remarkable.

ACKNOWLEDGEMENT

The author wants to express his gratitude for valuable information and advice from the following individuals: E. Sohl and G. D. Andrews (DuPont), S. E. Horne, Jr. (Polysar), R. T. Morrissey (Goodrich, Ret.), and J. V. Fusco, B. S. Henson, E. N. Kresge, and L. V. Mullen, Jr. (Exxon).

Elastomers Technology Division
Exxon Chemical Company
Linden, N.J. 07036

REFERENCES

1. Butlerov, H. A., and Goryainov, B., Ber. 6, 561 (1873)

2. Otto, M., Brennst.-Chem., 8, 321 (1927)

3. Tornqvist, E. G. M., 'The Historical Background of Synthetic Elastomers with Particular Emphasis on the Early Period', in Polymer Chemistry of Synthetic Elastomers, Pt. I., J. P. Kennedy and E. G. M. Tornqvist, Eds., Interscience-Wiley, New York, 1968, p. 21

4. Staudinger, H., Ber., 53, 1073 (1920)

5. Staudinger, H. and Fritschi, J., Helv. Chim. Acta, 5, 785 (1922)

6. Staudinger, H., Ber., 57, 1203 (1924)

7. Otto, M. and Mueller-Cunradi, M., U. S. Patent 2,130,507 (to I. G. Farbenindustrie), Sept. 20, 1938, convention date July 25, 1931

8. Otto, M. and Mueller-Cunradi, M., U. S. Patent 2,084,501 (to I. G. Farbenindustrie), June 22, 1937, convention date July 25, 1931

9. Howard, F. A. *Buna Rubber*, *The Birth of an Industry*, D. Van Nostrand, New York, 1947

10. Thomas, R. M., *Rubber Chem Technol.*, 42, G90 (1969)

11. Bannon, L. A., U. S. Patent 2,317,878 (to Jasco[3]), Apr. 27, 1943, filed Oct. 10, 1936

12. Otto, M., Gueterbork, H., and Hellemans, A., U. S. Patent 2,311,567 (to Jasco), Feb. 16, 1943, conv. Mar. 11, 1937

13. British Patent 358,058 (by Bataafsche Petroleum Co.), Sept, 29, 1931

14. Thomas, R. M. and Slotterbeck, O. C., U. S. Patent 2,243,658 (to Standard Oil Development Co.) May 27, 1941, filed Oct. 2, 1937

15. Thomas, R. M. and Sparks, W. J., U. S. Patent 2,356,128 (to Jasco), Oct. 22, 1944, filed Oct. 20, 1939

16. Thomas, R. M., Lightbown, I. E., Sparks, W. J., Frolich, P. K., and Murphree, E. V., *Ind. Eng. Chem.*, 32, 1283 (1940)

17. Lightbown, I. E., Sparks, W. J., and Thomas, R. M., U. S. Patent 2,494,766 (to Jasco), Jan 17, 1950, filed Dec. 17, 1947

18. Sparks, W. J., Rubber Chem Technol., 36, xxiii (1963)

19. Thomas, R. M. and Lightbown, I. E., U. S. Patent 2,322,972 (to Jasco), June 15, 1943, filed July 1, 1939

---

3 The Joint American Study Company, organized in 1930 by Standard Oil Co. (N.J.) and I. G. Farben.

20. Thomas, R. M. and Lightbown, I. E., U. S. Patent 2,276,893 (to Standard Oil Development Co.), Mar. 17, 1942, filed Dec. 31, 1938

21. Crawford, R. A. and Morrissey, R. T., U. S. Patent 2,631,984 (to Goodrich), Mar. 17, 1953, filed Apr. 18, 1950

22. Crawford, R. A. and Morrissey, R. T., U. S. Patent 2,681,899 (to Goodrich), Jun. 22, 1954, filed Sep. 5, 1952

23. Morrissey, R. T. and Weiss, H. J., U. S. Patent 2,698,041 (to Goodrich), Dec. 28, 1954, filed June 30, 1951

24. Morrissey, R. T., Ind. Eng. Chem, **47**, 1562 (1955)

25. Morrissey, R. T. and Frederick, M. R., U. S. Patent 2,732,354 (to Goodrich), Jan. 24, 1956, filed Dec. 5, 1952

26. Morrissey, R. T., Rubber World, **138**, 725 (1958)

27. Baldwin, F. P., Rubber Chem Technol., **52**, G77 (1979)

28. Baldwin, F. P. and Thomas, R. M., U. S. Patent 2,964,489 (to Esso Res. & Eng. Co.), Dec. 13, 1960, filed May 31, 1955

29. Morrissey, R. T., Personal communication

30. Baldwin, F. P., Buckley, D. J., Kuntz, I., and Robison, S. B., Rubber and Plastics Age, **42**, 500 (1961)

31. Van Tongerloo, A. and Vukov, R., Proc. Int. Rubber Conf., Milan, Italy, 1979, p. 70

32. Vukov, R., Rubber Chem. Technol., **57**, 275 (1984)

33. McQueen, D. M., U. S. Patent 2,212,786 (to DuPont), Aug. 27, 1940, filed May 1, 1939

34. McAlevy, A., Strain, D. E., and Chance, F. S., U. S. Patents 2,416,060 and 2,416,061 (to DuPont), Feb. 18, 1947, filed Oct. 27, 1946

35. McAlevy, A., U. S. Patent 2,586,363 (to DuPont), Feb. 19, 1952, filed May 19, 1947

36. Brooks, R. E., Strain, D. E., and McAlevy, A., India Rubber World, 127, 791 (1953)

37. Sohl, E. and Andrews, G. D., Personal communication

38. Ziegler, K., Holzkamp, E., Breil, H., and Martin, H., Angew. Chem., 67, 541 (1955)

39. Heuse, R. O., U. S. Patent 2,982,759 (to DuPont), May 2, 1961, filed Oct. 20, 1958

40. Ziegler, K., Intern. Symp. Macromol. Chem., Prague, 1957; Main and Section Lectures, Coll. Czech. Chem. Comm., 22, 295 (1957)

41. Natta, G., Pino, P., and Farina, M., Proc. Intern. Symp. Macromol. Chem. Milan, 1954; Ric. Sci. Suppl. A, 1955, pp. 120-133

42. Natta, G., and Porri, L., 'Diene Elastomers', in Polymer Chemistry of Synthetic Elastomers, Part II, Kennedy, J. P. and Tornqvist, E. G. M., Eds., Interscience-Wiley, New York, 1969, p. 597

43. Natta, G., Valvassori, A., and Sartori, G., 'Ethylene-Propylene Rubbers', in Polymer Chemistry of Synthetic Elastomers, Part II, Kennedy, J. P. and Tornqvist, E. G. M., Eds., Interscience-Wiley, New York, 1969, p. 679

44. Ziegler, K., Angew. Chem., 76, 545 (1964)

45. Hercules Chemist, No. 46, Feb. 1963, p. 7 (as quoted by Ziegler, Ref. 44)

46. Bier, G., Angew. Chem., 73, 186 (1961)

47. Natta, G., Rubber and Plastics Age, 38, 495 (1957); Chim. Ind. (Milan), 39, 653 (1957)

48. Natta, G., Mazzanti, G., Valvassori, A., and Pajaro, G., Chim. Ind. (Milan), 39, 733 (1957)

49. Mazzanti, G., Valvassori, A. , and Pajaro, G., Chim. Ind. (Milan), 39, 743 (1957)

50. Mazzanti, G., Valvassori, A., and Pajaro, G., Chim. Ind. (Milan), 39, 825 (1957)

51. Natta, G., Valvassori, A., Mazzanti, G., and Sartori, G., Chim. Ind. (Milan), 40, 717 (1958)

52. Natta, G., Valvassori, A., Mazzanti, G., and Sartori, G., Chim. Ind. (Milan), 40, 896 (1958)

53. Baldwin, F. P. and Ver Strate, G., Rubber Chem. Technol., 45, 709 (1972)

54. Gresham, W. F. and Hunt, M., U.S. Patent 2,933,480 (to DuPont), April 19, 1960, filed January 25, 1956

55. Anderson, A. W. and Merckling, N. G., U.S. Patent 2,721,198 (to DuPont), October 18, 1955, filed August 30, 1954

56. Anderson, A. W., Merckling, N.G., and Settlage, P. H., U.S. Patent 2,799,668 (to DuPont), July 16, 1957, filed September 19, 1956

57. Gladding, E. K., Fisher, B.S., and Colette, J. W., Ind. Eng. Chem., Prod. Res. Develop., 1, 65 (1962)

Francesco Ciardelli and Carlo Carlini

OPTICALLY ACTIVE POLYOLEFINS

1. INTRODUCTION

Optical activity can be observed in organic molecules provided conformers with a single chirality prevail on those having opposite absolute configuration. This occurs in general because of the presence of asymmetric carbon atoms in the molecule itself, purely conformational optical activity being limited to special cases(atropoisomers and so on), where rotation about bonds is hindered by structural reasons. Accordingly, even if polyethylene chains can assume chiral conformations, no way seems to exist allowing separation of structures with opposite chirality. In the case of 1-olefins, such as propylene, each monomeric unit contains an asymmetric carbon atom and then in principle the possibility exists of obtaining optically active polymers. However atactic and syndiotactic macromolecules can be considered non dissymmetric because of substantial intramolecular compensation inherent to their molecular structure(1). In isotactic chains, which by definition are formed by asymmetric carbon atoms all with the same absolute configuration, intramolecular compensation can also occur as the asymmetry is due to the different length of the two main chain sections attached to a single tertiary carbon atom. Indeed when the two end-groups are identical, isotactic chains have a pure "meso" structure(1). In any case in the presence of conventional polymerization catalysts no optically active polymers can be obtained as no reason exists for the predominance of a single chirality, independently of stereoregularity type and degree. No attempts are reported concerning the separation of polymers of 1-olefins in fractions with opposite rotatory power. On the other side optically active polymers of propylene resulted from the polymerization with catalysts obtained by reacting $TiCl_4$ with tris-[(S)-2-methylbutyl]-aluminum•etherate, but the specific rotatory power was low and decreased with increasing molecular weight, being not detectable for $\overline{M}_v$ larger than 100,000 (Table 1); this indicates that the observed optical activity was

R. B. Seymour and T. Cheng (eds.), History of Polyolefins, 163–176.

due only to the chiral end-groups derived from the catalyst(2).

TABLE 1. Rotatory power of polypropylenes obtained by $TiCl_4$/tris[(S)-2-methylbutyl]aluminum·etherate

Sample	$\overline{M_v}$	$\|\alpha\|_D^{20}$
1	750	+ 0.9
2	10,000	+ 0,4
3	180,000	+ 0.0
4	270,000	+ 0.0

This suggested that, in order to prepare optically active polyolefins, it was then necessary to use chiral olefins, which provide the presence of at least one asymmetric carbon atom in the side chain of each monomeric unit. The configuration of such asymmetric center is independent of molecular weight and stereoregularity, thus optical activity can be observed in the polyolefin provided one absolute configuration is predominant.

2. POLYMERIZATION OF OPTICALLY ACTIVE 1-OLEFINS

2.1. Polymerization of practically pure enantiomers

The first attempt to prepare optically active polyolefins was based on the polymerization of a single enantiomer(3-5). The capacity of these polymers to rotate the plane of polarized light excluded complete racemization, but was not a definite proof of complete retention. It was then of interest to determine the extent of possible inversion at the monomer asymmetric center. This evaluation was performed by controlling the optical purity of the non polymerized monomer at different conversions as poly-1-olefins, due to their paraffinic structure, can be cleaved only under very drastic conditions. It was shown that the unreacted monomer had optical purity equal or slightly lower than the starting monomer, small differences being related to distance of asymmetric carbon atom from the double bond, conversion and initial enantiomeric prevalence(6). This observation was later further substantiated by ^{13}C-NMR on polymers of optically active 1-olefins(7).
The absence of racemization during monomer insertion in the

growing chain is also in keeping with the independence of rotatory power from polymerization degree when this last is large enough(8).
In addition to configurational regularity in the side chains (enantiomeric purity), also stereoregularity can strongly affect polymer rotatory power. Thus macromolecules having different stereoregularity (Scheme 1) and equal enantiomeric purity can be expected to have different chiroptical properties due to the different stereochemical arrangement in the main chain whereas asymmetric carbon atoms in the side chains have the same absolute configuration in both stereoisomers.

Scheme 1

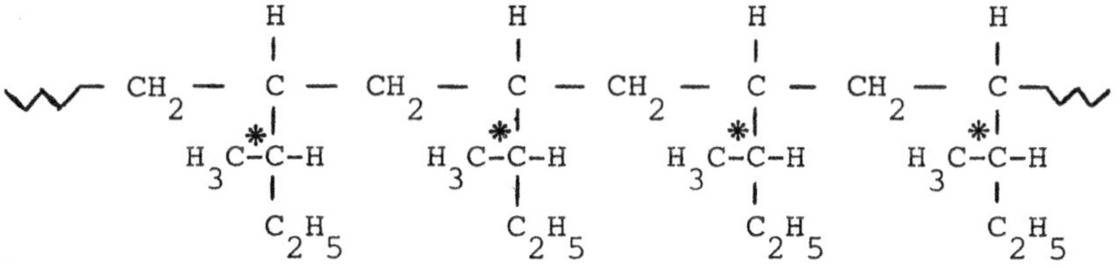

isotactic poly-(S)-3-methyl-1-pentene

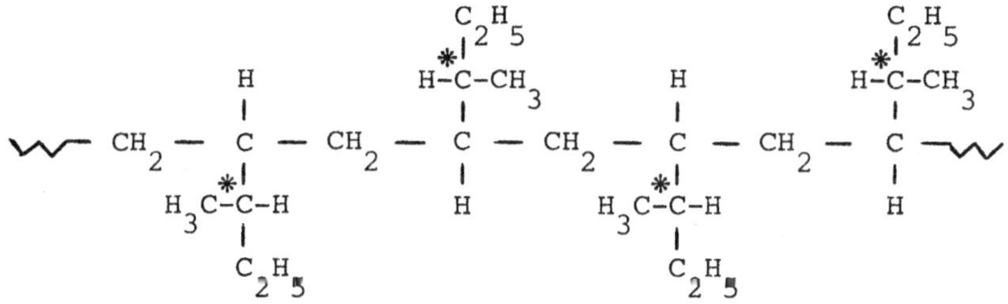

syndiotactic poly-(S)-3-methyl-1-pentene

This is particularly evident when the asymmetric carbon atom of the side chain is in α or β position to the main chain(9), where an increase of isotacticity degree gives an increase of absolute value of optical activity(Table 2).

2.2. Polymerization of non equimolar mixtures of enantiomers

Polymers prepared by polymerization of monomer samples having optical purity from 10 up to 95% show that, when the asymmetric carbon atom of the side chain is in the α or β position, no linear relationship exists between chiroptical properties and enantiomeric purity(10).

TABLE 2. Molar optical rotation of some optically active poly-1-olefins having different stereoregularity degree

1-olefin[a)]	fraction extracted successively with:	$[\eta]$[b)] (dl/g)	m.p. (°C)	$[\Phi]_D^{25}$ [c)]
(S)-3-methyl-	diethylether	0.13	93 - 96	+ 127
-1-pentene	isooctane	0.13	187 - 193	+ 146
	decalin	0.50	228 - 232	+ 161
(R)-3,7-dime-	acetone	n.d.	amorphous	- 45
thyl-1-octene	diethylether	0.26	n.d.	- 84
	isooctane	11.6	315 - 325	- 130
(S)-4-methyl-	acetone	n.d.	amorphous	+ 149
-1-hexene	ethylacetate	0.20	130 - 135	+ 239
	diethylether	1.29	181 - 186	+ 259
	diisopropylether	1.35	210 - 215	+ 279
(S)-5-methyl-	acetone	n.d.	amorphous	+ 27
-1-heptene	diethylether	5.8	54 - 55	+ 68

a) Optical purity: 89 - 95%; b) In decalin at 120°C;
c) In hydrocarbon solution, based on one monomeric unit.

By contrast a linear relationship is observed when the above asymmetric carbon atom is in the γ position or further(10). Whereas, as discussed in section 3.1., a very modest preferential polymerization of the predominant enantiomer is observed(6) with monomers having the asymmetric carbon atom directly bound to the double bond, that is in α position, no appreciable variation of polymerized monomer optical purity seems to exist in other cases and particularly when in β. Indeed preferential polymerization of one antipode has been also excluded by using ^{14}C labelled 4-methyl-1-hexene and cecking radioactivity of starting monomer against that of polymer and recovered nonpolymerized monomer(6) (Scheme 2). It is of interest to remark that, when no linear relation exists, the maximum rotatory power was observed with monomer optical purity about 70-80%. This result was explained admitting a certain degree of copolymerization between the two antipodes, which is probably larger when the asymmetric carbon atom is in the β rather than in the α position(11).

Scheme 2

Polymerization of (S)-4-methyl-1-hexene-1-^{14}C/(R)-4-methyl-1-hexene 3/1 mixture(6).

Starting monomer (radioactivity 36900 cpm) —$TiCl_4/Al(i.Bu)_3$→ polymer (radioactivity 36500 cpm)

→ recovered nonpolymerized monomer (radioactivity 36700 cpm)

In the copolymer macromolecules the units derived from the predominant antipode force the units with opposite configuration, present in lower concentration, to assume conformations giving optical activity of the same sign as the former units.

3. POLYMERIZATION OF RACEMIC 1-OLEFINS

3.1. Stereoelective polymerization

Attempts to prepare optically active polymers with racemic monomers and optically active catalysts led to the preferential polymerization of one antipode from the racemic mixture (Scheme 3)(12).

Scheme 3

racemic monomer (S + R) —optically active (S)-catalyst→ optically active (S)-polymer + optically active (R)-monomer

This polymerization was called <u>stereoelective</u>(13), according to the proposal that stereoelective polymerization occurs when from a mixture of stereoisomers, each of which can be the origin of an unique type of monomeric unit, one obtains a tactic polymer(14). Catalytic systems prepared both by reacting a transition metal derivative with suitable optically active metal-alkyls which do not undergo rapid racemization reactions(13), and by addition of an optically active third component to the conventional Ziegler-Natta catalyst(15) resulted to be effective in inducing stereoelective polymerization, as shown in Tables 3 and 4, respectively.

TABLE 3. Stereoelective polymerization of racemic 3,7-dimethyl-1-octene by $TiCl_4$ and different (S)-2-methylbutyl metal-alkyls(MeR^*_n) (13,16)

MeR^*_n a)	molar ratio MeR_n/Ti	conversion (C) (%)	recovered monomer optical purity(P_m) (%)	polymerized b) monomer optical purity(P_p) (%)
GaR_3	3	31.0	0.21	0.5
AlR_3	2	35.2	0.62	1.1
InR_3	3	35.2	0.69	1.3
$AlR_3 \cdot Et_2O$	3	7.9	0.15	1.7
ZnR_2	6	41.9	2.90	4.1
ZnR_2	3	26.9	3.86	10.5
ZnR_2	1	12.3	1.18	8.4

a) Optical purity 85-99%; b) Evaluated as: $P_m(100 - C)/C$

TABLE 4. Polymerization of racemic 3,7-dimethyl-1-octene by chirally modified Ziegler-Natta catalysts based on $TiCl_4$ and $Al(i.Bu)_3$ (15)

Chiral component in the catalyst	Polymeric product: absolute configuration	$[\alpha]^{25}_D$	polymerized monomer optical purity, (%)
(-)-α-pinene	R	- 3.4	∿ 2
(-)-menthyl-acetate	S	+ 9.0	∿ 5
(+)-α-phenyl-ethylamine	R	- 6.4	∿ 3
poly(-)-menthyl-vinylether	R	n.d.	∿ 11

Indeed, by interrupting the polymerization at conversion not higher than 50%, both recovered nonpolymerized monomer and polymer resulted to be optically active. The antipode prevailing in the final monomer had obviously opposite absolute configuration than that of the preferentially polymerized monomer(Scheme 3). The enantiomeric purity of the polymer was however never higher than 10% and no better results were obtained by changing structure of metal-alkyl or varying conversion(16). Fractions with opposite optical rotation related to different tacticity were obtained from polymers of racemic 4-methyl-1-hexene prepared by supported Ziegler-Natta catalyst chirally modified with (-)-menthyl anisate(17).

3.2. Separation of polymers from racemic monomers into fractions having opposite optical rotation

Investigation of antipodes distribution in polymers from racemic 1-olefins started with clear mechanistic objectives in order to establish if, in the presence of Ziegler-Natta catalysts giving isotactic polymers, the two antipodes were yielding a random copolymer or a mixture of homopolymers (Scheme 4). The formation of a crystalline polymer from racemic 4-methyl-1-hexene suggested already in 1955 that the latter hypothesis was more probable(18). This was later proved as the

Scheme 4

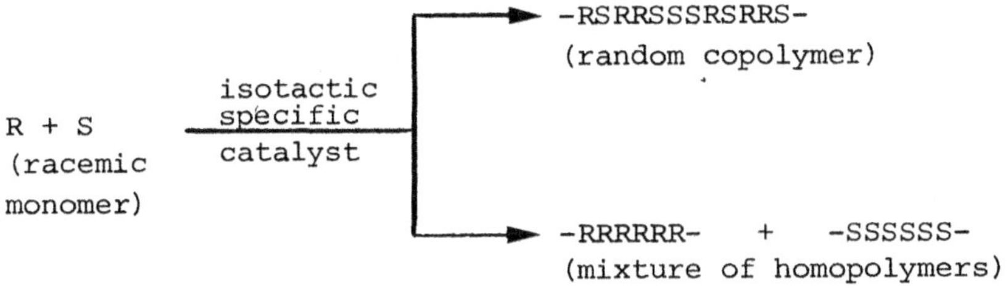

polymers from racemic monomers having the asymmetric carbon atom in the α or β position to the double bond were separated into fractions having opposite optical rotation by chromatography on an optically active support like isotactic crystalline poly-1-olefins(19) (Table 5). This separation was accompanied by fractionation according to molecular weight and stereoregularity, thus strongly reducing separation efficiency. By evaluating the efficiency degree, it was successively demonstrated(20) that, at least when the asymmetric carbon atom is in the α position to the double bond, the two enantiomeric monomers are separately inserted in different growing chains, giving rise to a mixture of homopolymers of the two antipodes (Scheme 4) with stereoselectivity as high as 90-95% (<u>Stereoselective polymerization</u>) (21) (Table 6).
This conclusion was furtherly supported copolymerizing a racemic monomer with an optically active analog in the presence of a conventional stereospecific catalyst(22).

TABLE 5. Separation of isotactic polymers from racemic monomers in fractions with opposite rotatory power by elution on poly(S)-3-methyl-1-pentene(19)

Polymer from racemic:	optically active polymer fractions showing the largest rotatory power			
	wt.%	$[\alpha]_D^{25}$	predominant absolute configuration	approximate enantiomeric purity, %
3-methyl-1--pentene	11.2	- 26	R	16
	11.7	+ 40	S	25
3,7-dimethyl--1-octene	12.8	+ 30	S	23
	7.7	- 60	R	47
4-methyl-1--hexene	12.1	- 24	R	8
	7.8	+ 98	S	34

TABLE 6. Intrinsic separability in fractions of opposite optical rotation of polymers from racemic 3,7-dimethyl-1-octene (20).

Sample submitted to elution on poly-(S)--3-methyl-1-pentene	separation degree,% F	efficiency E	intrinsic separability,% D = F/E
Equimolar mixture of poly-(R)- and poly-(S)--3,7-dimethyl-1-octene	37.0	0.37	100
poly-(R)(S)-3,7-dimethyl--1-octene	33.6	0.37	91

The main requisite of stereoselective catalysts is the capability of the active sites to distinguish between the two enantiomorphic forms of the monomer(11). As a consequence the two monomers with the same configuration give rise to copolymer macromolecules while the remaining monomer yields its homopolymer(Scheme 5).

Scheme 5

R + S + S' $\longrightarrow$ -RRRRR- + -SS'SSSS'S-

R + S	+ S' $\longrightarrow$	-RRRRR- +	-SS'SSSS'S-
racemic monomer	S-monomer	R-homopolymer (less soluble)	(S)-copolymer (more soluble)

Therefore solvent extraction of the polymeric products gives first copolymers fractions with, for instance, positive optical rotation and finally homopolymer with opposite(negative) optical rotation.

4. COPOLYMERS OF OPTICALLY ACTIVE 1-OLEFINS WITH ACHIRAL OLEFINS

In the cases mentioned in sections 2 and 3 the polymer optical activity derives from configurational predominance in the side chain, while stereoregularity and conformational properties, which are interconnected, affect the order of magnitude of optical rotation. Indeed, the absolute value of molar rotatory power of polyolefins varies markedly with isotacticity degree(Table 2), distance of asymmetric carbon atom from backbone, and length of isotactic blocks from the chiral monomer. The last two effects are clearly shown by some typical examples reported in Table 7, indicating that rotatory power of the most isotactic fraction decreases in absolute value with moving away side chain chiral center farther than α-position from backbone and by insertion of ethylene units in the isotactic polymer chain of an optically active monomer.

TABLE 7. Molar rotatory power of isotactic poly-1-olefins and of copolymers with ethylene

Polymer	$[\Phi]_D^{25}$ per chiral unit	molar fraction of chiral units in the favored screw sense	Ref.
poly-(S)-3-methyl-1--pentene	+ 161	∿ 0.95	9,25
poly-(S)-4-methyl-1--hexene	+ 228	∿ 0.95	9,25
poly-(S)-5-methyl-1--heptene	+ 68	∿ 0.70	9,25
poly-(S)-6-methyl-1--octene	+ 24	∿ 0.55	9,25
ethylene/(S)-4-methyl--1-hexene(50/50) copolymer	+ 133	∿ 0.70	23

The optical rotatory properties of isotactic polyolefins were interpreted on the basis of the presence of conformations in solution of the same type as observed in the crystalline state(24). When chiral groups of a predominant configuration are not present both screw senses have equal statistical weight, while the polymeric product is able to rotate the plane of polarized light if a single screw sense prevails. Such a prevalence is remarkable when the polymer is isotactic and the asymmetric center of a single chirality in the side chain is in the α or β position(Table 7) to the main chain, cooperative effects along blocks being determinant to have practically one screw sense only(25). On the other side the allowed conformations for the monomeric unit in an isotactic polymer derived from an achiral vinyl monomer belong either to the left handed or to the right handed helix, each conformation having the same statistical weight as its mirror image. In the case of 4-methyl-1-pentene for instance, only one pair of enantiomeric conformations is allowed, the respective molar optical rotation calculated according Brewster(26) being - 240 and + 240 at the sodium-D-line. A polymer where one of the above conformations prevails over its mirror image has been obtained by copolymerizing 4-methyl-1-pentene with the chiral analog (S)-4-methyl-1-hexene in the presence of stereospecific isotactic Ziegler-Natta catalyst(27). The rotatory power of the copolymer is higher in absolute value than that of the homopolymers mixture whatever the composition. From the difference in optical rotation between copolymer and homopolymers mixture having the same composition, the molar optical rotation referred to one monomer residue of 4-methyl-1-pentene was evaluated. This last increased in absolute value with increasing the content of units derived from the optically active comonomer and reached an asymptotic value when the above content is larger than 50%. The optical rotation of 4-methyl-1-pentene units thus obtained can be of the same order of magnitude as that evaluated according to Brewster, assuming that these units are inserted in one screw sense helical chain and corresponds to a content of more than 80% units inserted in left-handed helical conformations(27).

5. FINAL REMARKS

The investigation concerning optically active 1-olefins has given two main contributions to the understanding of this important class of polymers.

Indeed the preparation of optically active polymers by polymerization of enantiomers of chiral 1-olefins(6) and by their copolymerization with nonchiral 1-olefins has provided an evident proof of the permanence in solution of chain conformations consisting of more or less long blocks with one screw sense helical secondary structure connected through conformational reversals(9,25). The existence of this ordered structures is connected with steric cooperative interactions between chiral side chains as no other type of interactions exists due to the paraffinic nature of the polymer. Helical sections content changes regularly with temperature because of rapid conformational interconversion and then no conformational transition can be observed. In optically active polyolefins one screw sense prevails on the opposite one because of the single chirality of the side chain; this effect can be transmitted to sections of nonchiral 1-olefins provided they are included with optically active comonomers in the same macromolecule(27).
The use of chiral monomers has brought a second important contribution to the understanding of stereochemistry of Ziegler-Natta catalysts. Indeed the separability of polymers from racemic monomers into fractions of opposite sign rotatory power up to more than 90% is to be considered one of the strongest support to the intrinsic dissymmetry of active sites, excluding important steric control by the growing chain-end(20). This conclusion was furtherly confirmed by showing that the random insertion of ethylene units in the growing chain did not affect the degree of separability D in fractions with opposite rotatory power(28).
At present optically active polyolefins cannot find practical applications as commodities. Indeed these materials are still rather expensive; copolymerization of pure enantiomers with non chiral ready available comonomers may however help to overcome this drawback(29). In any case it cannot be excluded that the present trend, focused on speciality polymers, can benefit of these very special polymers in area where chemical stability and molecular dissymmetry are requested.

Centro di Studio del C.N.R. per le Macromolecole Stereordinate ed Otticamente Attive

Dipartimento di Chimica e Chimica Industriale, Università di Pisa, 56100 Pisa, Italy.

REFERENCES

1. P. Pino, 'Optically active addition polymers', *Adv. Polym. Sci.*, 4, 393(1965)
2. P. Pino, F. Ciardelli, G.P. Lorenzi, 'Optically active vinyl polymers. VIII. Some aspects of the stereospecific polymerization of racemic α-olefins', *J. Polym. Sci., C*, 4, 21(1964)
3. P. Pino, G.P. Lorenzi, L. Lardicci, 'Optically active vinyl polymers. I. Preparation of isotactic polymers of optically active α-olefins', *Chim. Ind.(Milan)*, 42, 712(1960)
4. W.J. Bailey, E.T. Yates, 'Polymers.III. Synthesis of optically active stereoregular polyolefins', *J. Org. Chem.*, 25, 1800(1960)
5. S. Nozakura, S. Takeuchi, H. Yuki, S. Murahashi, 'Stereoregular polymers.IX. Polymerization of optically active α-olefins', *Bull. Chem. Soc. Japan*, 34, 1673(1961)
6. F. Ciardelli, G. Montagnoli, D. Pini, O. Pieroni, C. Carlini, E. Benedetti, 'Stereospecific polymerization of asymmetric α-olefins having different optical purity', *Makromol. Chem.*, 147, 53(1971)
7. F. Conti, L. Acquaviva, E. Chiellini, F. Ciardelli, M. Delfini, A.L. Segre, '^{13}C.n.m.r. study of optically active polymers: poly(4-methyl-1-hexene)', *Polymer*, 17, 901 (1976)
8. P.L. Luisi, F. Pezzana, 'Fractionation of stereoregular poly[(S)-4-methyl-1-hexene] and poly[(R)(S)-4-methyl-1--hexene] on the basis of molecular weight', *Europ. Polym. J.*, 6, 259(1970)
9. P. Pino, F. Ciardelli, G.P. Lorenzi, G. Montagnoli, 'Optically active vinyl polymers.IX. Optical activity and conformation in dilute solution of isotactic poly-α-olefins', *Makromol. Chem.*, 61, 207(1963)
10. P. Pino, F. Ciardelli, G. Montagnoli, O. Pieroni, 'On the relationship between monomer optical purity and polymer rotatory power in some linear poly-α-olefines', *J. Polym. Sci., B*, 5, 307(1967)
11. P. Pino, A. Oschwald, F. Ciardelli, C. Carlini, E. Chiellini, 'Stereoselection and stereoelection in α-olefin polymerization', in *Coordination Polymerization: A Memorial to Karl Ziegler*, J.C.W. Chien Ed., Academic Press, New York, 1975, p. 25
12. P. Pino, F. Ciardelli, G.P. Lorenzi, 'Optically active vinyl polymers.X. Synthesis of optically active poly-α-olefins from racemic monomers by asymmetric catalysts', *J. Am. Chem. Soc.*, 85, 3888(1963)

13. F. Ciardelli, C. Carlini, G. Montagnoli, L. Lardicci, P. Pino, 'Optically active vinyl polymers.XIX. Polymerization of racemic α-olefins in the presence of catalysts prepared by reaction of titanium tetrachloride with various optically active organometallic compounds', *Chim. Ind.(Milan)*, 50, 860(1968)
14. G. Natta, F. Danusso, 'Nomenclature relating to polymers having sterically ordered structure', *J. Polym. Sci.*, 34, 3(1959)
15a. R. Nocci, C. Carlini, F. Ciardelli, 'Stereoelective polymerization of racemic α-olefins by chirally modified Ziegler-Natta catalysts', *J. Polym. Sci., Polym. Chem. Ed.*, 15, 767(1977)
b. C. Carlini, F. Ciardelli, 'Polymerization of racemic α-olefins by Ziegler-Natta catalysts in the presence of optically active compounds', *Chim. Ind.(Milan)*, 63, 486 (1981)
16. C. Carlini, H. Bano, E. Chiellini, 'Stereoelective polymerization of racemic α-olefins: influence of conversion and of nature of catalyst components', *J. Polym. Sci., A-1*, 10, 2803(1972)
17. P. Pino, G. Fochi, O. Piccolo, U. Giannini, 'Experimental evidence for the existence of catalytic sites with different steric structure in supported Ziegler-Natta catalysts', *J. Am. Chem. Soc.*, 104, 7381(1982)
18. G. Natta, P. Pino, G. Mazzanti, P. Corradini, U. Giannini, 'Synthesis and properties of highly polymerized crystals of some α-olefins having branched chains', *Rend. Acc. Naz. Lincei[VIII]*, 19, 397(1955)
19a. P. Pino, F. Ciardelli, G.P. Lorenzi, G. Natta, 'Optically active vinyl polymers.VI. Chromatographic resolution of linear polymers of (R)(S)-4-methyl-1-hexene', *J. Am. Chem. Soc.*, 84, 1487(1962)
b. P. Pino, G. Montagnoli, F. Ciardelli, E. Benedetti, 'Optically active vinyl polymers.XVI. Some aspects of the chromatographic separation of linear polymers of racemic α-olefins in fractions having optical activity of opposite sign', *Makromol. Chem.*, 93, 158(1966)
20. G. Montagnoli, D. Pini, A. Lucherini, F. Ciardelli, P. Pino, 'Quantitative aspects of the stereoselectivity in the stereoregulated polymerization of racemic α-olefins by Ziegler-Natta catalysts', *Macromolecules*, 2, 684(1969)
21. M.L. Huggins, G. Natta, V. Desreux, H. Mark, 'Report on nomenclature dealing with steric regularity in high polymers', *J. Polym. Sci.*, 56, 153(1962)

22. F. Ciardelli, C. Carlini, G. Montagnoli, 'Stereochemical aspects of the copolymerization of asymmetric α-olefins by stereospecific catalysts', *Macromolecules*, 2, 296 (1969)
23. O. Pieroni, G. Stigliani, F. Ciardelli, 'Copolymerization of ethylene with asymmetric α-olefins: stereoselectivity of the process and optical activity of the copolymers', *Chim. Ind. (Milan)*, 52, 289 (1970)
24. G. Natta, P. Corradini, 'Structure of crystalline polyhydrocarbons', *Suppl. Nuovo Cimento*, 15, 9 (1960)
25. P.L. Luisi, F. Ciardelli, 'Configuration and conformation of high polymers', in *Reactivity, Mechanism and Structure in Polymer Chemistry*, A.D. Jenkins, A. Ledwith Eds., J. Wiley & Sons, New York, 1974, p. 471
26. J.H. Brewster, 'An useful model of optical activity. I. Open chain compounds', *J. Am. Chem. Soc.*, 81, 5475 (1959)
27. C. Carlini, F. Ciardelli, P. Pino, 'Optical activity and conformation in solution of stereoregular copolymers of (S)-4-methyl-1-hexene with 4-methyl-1-pentene', *Makromol. Chem.*, 119, 244 (1968).
28. F. Ciardelli, P. Locatelli, M. Marchetti, A. Zambelli, 'Stereoselective copolymerization of racemic α-olefins with ethylene by isospecific Ziegler-Natta catalysts', *Makromol. Chem.*, 175, 923 (1974)
29. F. Ciardelli, M. Aglietto, C. Carlini, E. Chiellini, R. Solaro, 'Structural analysis of optically active copolymers by CD', *Pure Appl. Chem.*, 54, 521 (1982)

Acknowledgement

Partial support by M.P.I. (Sci. Res. 60%) is gratefully acknowledged.

R. J. Statz

ETHYLENE COPOLYMER IONOMERS

A discussion of the discovery of ethylene copolymer ionomers is presented. The structural features of ionomers and how these features are responsible for the physical properties of ionomers is given. Importance of these properties in various end use applications is then demonstrated. Finally, recent developments in Surlyn® ionomer technology are discussed.

INTRODUCTION

In the 1940's Du Pont and other companies investigated and began producing copolymers. In general, adding comonomers to the polymerization reduces the ethylene crystallinity, thereby reducing the melting point, freezing point, the tensile strength and modulus. At the same time, optical properties were improved and polarity was increased. Copolymers first became commercially available in the 1950's. Copolymer technology, which was one step away from the original high pressure polyethylene technology, led to many new products. By changing comonomer type and level, the industry developed copolymers which have found uses in variety of film, molding, and additive applications.

Ionomer technology can be thought of as the second step away from polyethylene technology. This second step is the controlled neutralization with a cation (sodium or zinc) of ethylene/methacrylic acid copolymers. Neutralization of the acid copolymer creates a three dimensional network in the ionomer. These ionic crosslinks are responsible for the major property differences between ionomers and ethylene copolymers.

R. B. Seymour and T. Cheng (eds.), History of Polyolefins, 177–192.

HISTORY

During the early 1960's a small group of chemists at Du Pont began investigating ethylene copolymers where the comonomers had functionality. This investigation led to the discovery of ethylene/methacrylic acid copolymers, their unique rheology and their improved adhesion to aluminum foil (Ref. 1). Dr. Richard W. Rees, a research scientist in this group, was investigating methods for chemically crosslinking ethylene copolymers. He wanted to avoid peroxides and radiation so he investigated chemical methods.

One of the routes he evaluated was the reaction of epichlorohydrin with the sodium salt of ethylene methacrylic acid copolymers. During the isolation of the intermediate sodium salt of the acid copolymer it was realized that the material had unique properties. It was clear as glass, very tough and resilient and could be processed similar to polyethylene.

These materials were first disclosed in a September issue of Modern Plastics (Ref. 2) and a composition of matter patent was granted to Du Pont in August, 1966 (Ref. 3).

A number of people at the Du Pont Experimental Station and at the Sabine River Works in Texas were able to develop a process and a number of products which led to the successful commercialization of these ionomeric resins which were called Surlyn®.

MODEL

A number of people at Du Pont and a large number of people in academic institutions have studied the structure of ethylene ionomers. (Ref. 5-14) Several elegant review articles describing ionomers structure have been written by R. Longworth at Du Pont (Ref. 15,17).

The following describes one theory of the structure of ionomers.

Figure I is a schematic representation of a typical ethylene methacrylic acid copolymer. It shows lamella crystallites aligned radially in a manner similar to the structure found in the crystalline regions of polyethylene. In addition to these lamella there is an amorphous region, which consists of a variety of chain defects, short chain branches, and carboxyl groups, which can be considered short chain branches.

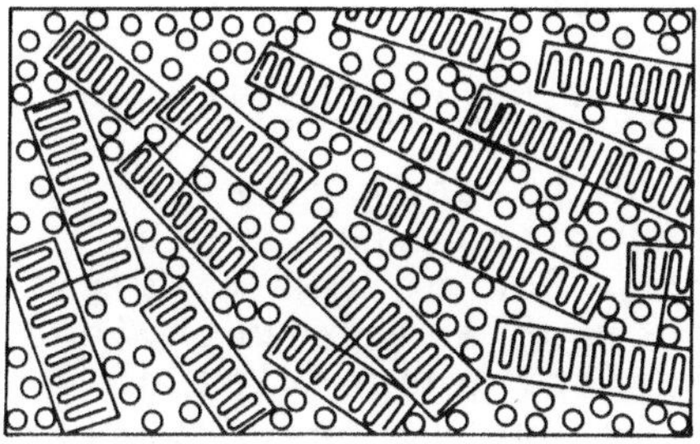

SCHEMATIC STRUCTURE OF E/MAA COPOLYMER

100Å

Figure II is a model of the ionomer based on the ionic cluster theory. This theory postulates that the ionic regions that are formed by neutralization of the carboxylates remove the acid groups from the amorphous phase and cause the crystalline regions to be less ordered. Evidence shows that the ionic phase undergoes a type of clustering. However, the exact structure of these clusters is subject to much controversy.

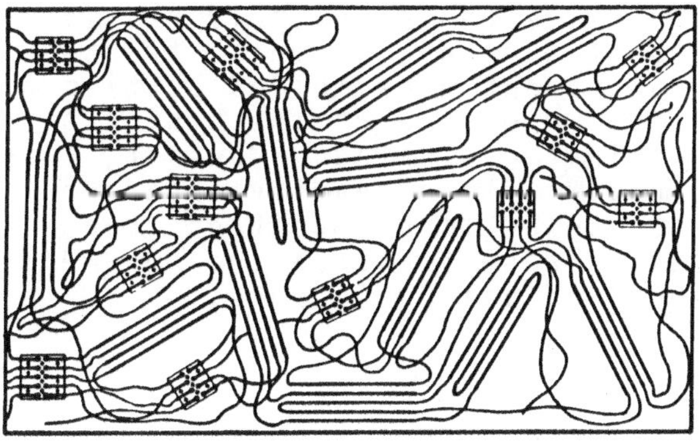

SCHEMATIC STRUCTURE OF IONOMER

100 Å (10 nm)

—o MAA Group

≡ Polyethylene Crystallite

\+ Cation

□ Ionic Cluster

Furthermore, the ionic cluster theory postulates the existence of three phases rather than the two phases found in ethylene methacrylic acid copolymers. The phases are crystalline polyethylene, ionic copolymer and amorphous (or branched) polyethylene which serves to connect the ionic phase with the crystalline regions.

COMMERCIALIZATION - PACKAGING PROPERTIES AND USES

No discussion of the history of a product can be complete without some discussion of the commercial implications. One of the most visible differences between ionomers and polyethylene is the greatly improved optical properties of ionomers. The elimination of large crystallites in the ionomer significantly reduces the internal haze from scattered light. High melt viscosities are responsible for hindering large crystallite growth. At the same time, nucleation by ionic regions results in the growth of small crystallites. Therefore, ionomers are more transparent in thick sections, and they show higher gloss and brilliance in film form than is shown by other ethylene polymers.

This very visible improvement in optical properties of ionomers provided the major incentive to commercialize these new polymers. Early market development of Surlyn® ionomers stressed clarity; hence transparent blow molded bottles and molded optical parts were some of the products in which Surlyn® was first evaluated. However, like many other new products, less readily apparent features turned out to be more important than optical properties. These features include melt strength, toughness and oil resistance.

High Melt Strength

The same ionic forces that suppress large crystallite growth and reduce haze, act on the molten polymer to create a highly viscous elastic melt that can be rapidly drawn to thin gauges without breaking. In end-use fabrication, the very viscous high melt strength ionomer controls the processing of blown film, extrusion coated substrates, skin and heat sealable packages, and deep draw vacuum containers.

MELT VISCOSITY OF 10% ACID IONOMER COMPARED TO LOW DENSITY POLYETHYLENE

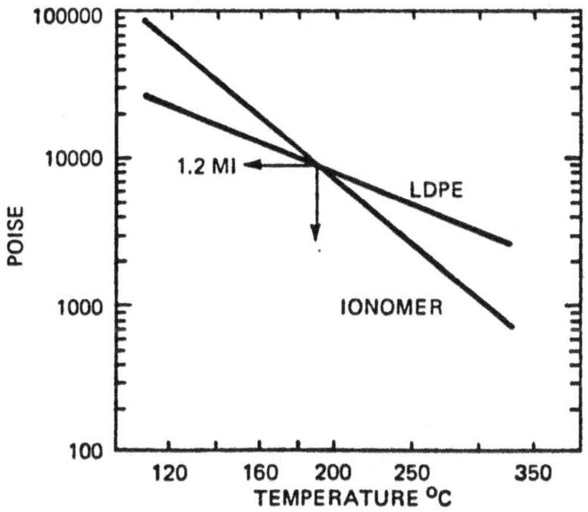

Graph I shows viscosity versus temperature for both an ionomer and low density polyethylene. In the two resins chosen for comparison, the ionomer's activation energy at constant shear rate is 10 Kcal/mol versus a typical 5 Kcal/mol for polyethylene. At lower temperatures, ionomers have much higher viscosities than LDPE. Therefore, they have excellent blown film bubble stability, and they are easily converted into foams.

Graph II showing draw rate vs melt tension highlights the fact that control of compositional variables offers a wide range of control over melt strength. The opportunity exists to tailor resins to specific processing requirements such as drawdown in extrusion coating. Furthermore, high melt strength is the factor behind the low incidence of pinholes found in thin coatings of ionomer on kraft paper. By contrast, the presence of fiber ends will puncture the molten web of polyethylene and other copolymers. The higher melt strength of ionomers is responsible for such puncture resistance.

DRAW RATE vs MELT TENSION

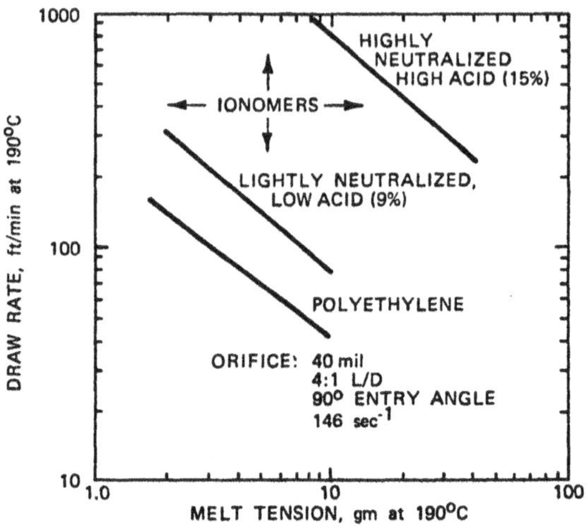

Draw rate also varies directly with temperature. This gives the user greater flexibility for either broader processing control, as in foamed extrusions, film production and extrusion coating or alternatively, faster operations for in-line packaging systems.

A part of the total mechanism involved in heat sealing is hot tack strength. Notice in Graph III that for a broad range of temperatures, Surlyn® provides substantially greater hot tack strength than do other ethylene polymers.

To summarize, melt strength can be related to heat seal, hot tack, deep draw vacuum forming, and puncture resistant molten webs. Commercial applications in which this property is important include form fill liquid packaging, skin packaging and film coated uses.

High Solid State Toughness

There is dramatic evidence that as we increase neutralization, or the ionic interchain forces, we increase the toughness of the acid co-

HOT TACK STRENGTH OF HEAT SEAL RESINS
vs INTERFACE SEAL TEMPERATURES

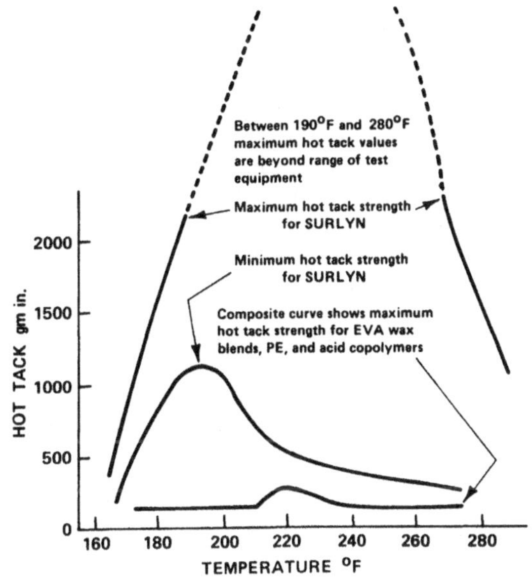

BLOWN FILM IMPACT STRENGTH
(2 mil FILM; 1.6:1 BUR)

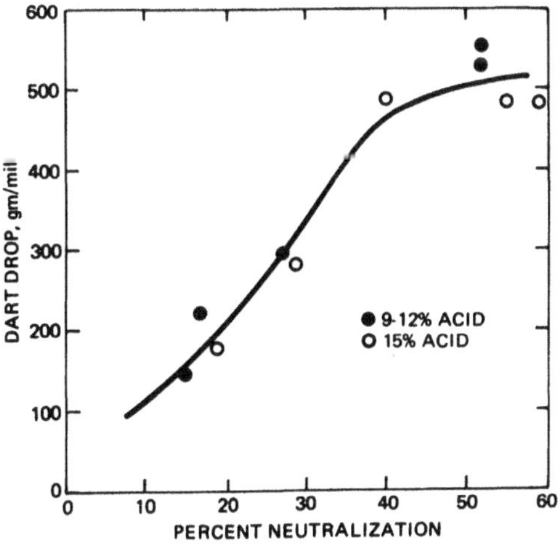

polymer matrix. Graph IV relates the blown film impact strength of ionomers to the percent neutralization. Surlyn® ionomers are often coated on paper and foil in uses where toughness is required.

In composite vacuum-packaged structure, the flexural crack resistance leads to fewer leaking packages and less spoilage. Figure III lists comparative data for nylon/ionomer and nylon/polyethylene composite films. The nylon/ionomer structure exhibits greater toughness as measured by the Dart Drop test and greater tear strength as measured by the Elmendorf Tear test.

COMPOSITE FILMS

	(1mil) NYLON (2 mils) IONOMER	(1 mil) NYLON (2 mils) POLYETHYLENE
DART DROP (60" HT.)	251 gms/mil	126 gms/mil
ELMENDORF TEAR MD/TD	65/360 gms/mil	25/200 gms/mil
HAZE	1.28%	6% (After Forming)
HEAT SEALS	280°F – 450°F +	350°F – 425°F

Oil Resistance

Ionomers display excellent resistance to permeation by edible oils. Figure IV compares the permeation of cottonseed oil and peanut oil through one-mil coatings of high density polyethylene, ethylene methacrylic acid copolymer and sodium ionomer derived from the same acid copolymer.

We know the introduction of polarity by inclusion of carboxylic acid groups in the amorphous segments of the copolymer reduces permeability to oil compared with low density polyethylene. Here we see that the highly polar ionic phase reduces permeability to oil versus highly crystalline linear polyethylene. Now the presence of ionized carboxyl groups by the addition of sodium cations to the acid copolymer dramatically illustrates the effect of the inter-

OIL PERMEABILITY OF IONOMERS

Test: Breakthrough on 1 mil coatings at 60 °C.

Polymer Sample	Cotton Seed Oil (Hours to Failure)	Peanut Oil (Hours to Failure)
High-Density Polyethylene (0.960 Density)	40	25
Ethylene-Methacrylic Acid (3.5 Mole % Acid)	64	78
Sodium Ionomer Derived From 3.5 Mole % Acid copolymer (55% Ionization)	229	182

chain ionization. Not surprisingly, it can be shown that this resistance to oil permeation extends over a broad range of temperatures.

MOLDING PROPERTIES AND USES

Ionomers have also found wide applications in the molding and extrusion industry. The properties which are of principle importance in this industry are stiffness and resilience, abrasion resistance, and solid state toughness. A comparison of ionomer and noncarboxylic acid ethylene copolymers is shown in Figure V. At a given comonomer content, ionomers exhibit higher tensile strength, higher flexural modulus, improved hardness, and better abrasion resistance than ethylene/vinyl acetate and ethylene/ethyl acrylate resins.

COMPARISON OF IONOMER PROPERTIES WITH ETHYLENE COPOLYMER PROPERTIES

Properties	Medium Modulus Ionomer	Ethylene/ Vinyl Acetate	Ethylene/ Ethyl Acrylate
Comonomer	4.2	4.2	4.7
Tensile Strength MPa	27.6	19.3	13.8
Flexural Modulus MPa	250	66	38
Tensile Impact MJ/m^2	1.0	0.46	1.1
Hardness Shore D	64	38	35
Abrasion Resistance NBS	410	160	-

STIFFNESS

The first key property ionomers bring to molding uses is stiffness. Graph V shows that an increase in stiffness of the acid copolymer occurs at neutrlization levels up to 30%. When an ionomer is compared to an ethylene vinyl acetate or ethylene acrylate copolymer of equivalent crystallinity, the modulus will be two to five times higher. The ionomer will also be more resilient.

STIFFNESS (TINIUS OLSEN) vs DEGREE OF IONIZATION (MONOBASIC ACID)

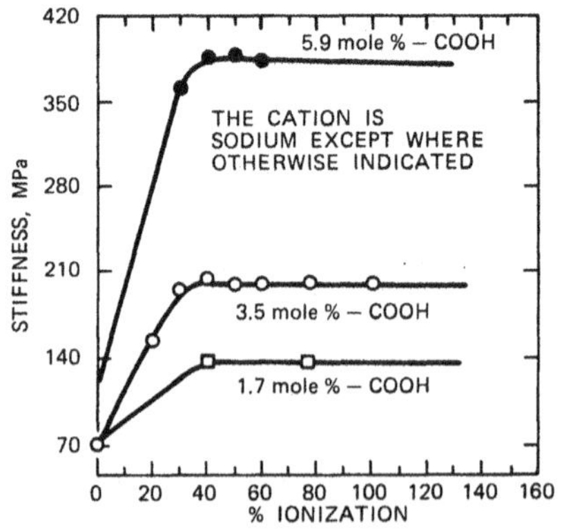

IMPROVED ABRASION RESISTANCE

Graph VI shows the increase in abrasion resistance as a 15% acid copolymer is neutralized. High abrasion resistance allows Surlyn® ionomers to compete with thermoplastic polyurethanes in applications such as ski boots and bowling pin covers.

ABRASION RESISTANCE vs % IONIZATION

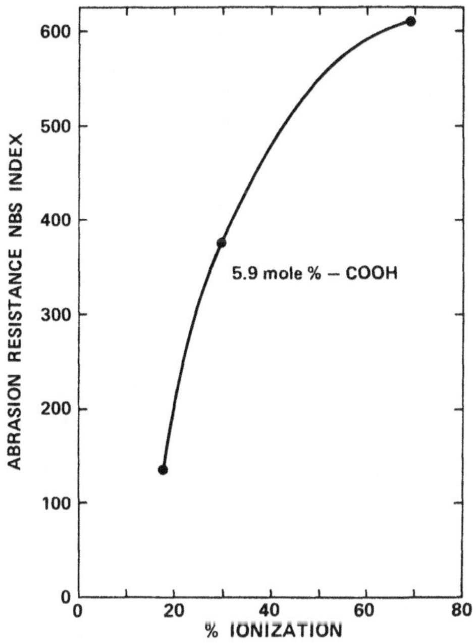

SOLID STATE TOUGHNESS

Like the increase in toughness observed for Surlyn® films, molded and extruded parts produced from ionomers have outstanding resistance to impact. Graph VII demonstrates this with a plot of tensile impact versus neutralization for a 10% acid copolymer. The lower line shows that at temperatures as low as -40°C, ionomers maintain their high solid state toughness. Because of this toughness, Surlyn® has found use in golf ball shells, automotive bumper guards, and elsewhere.

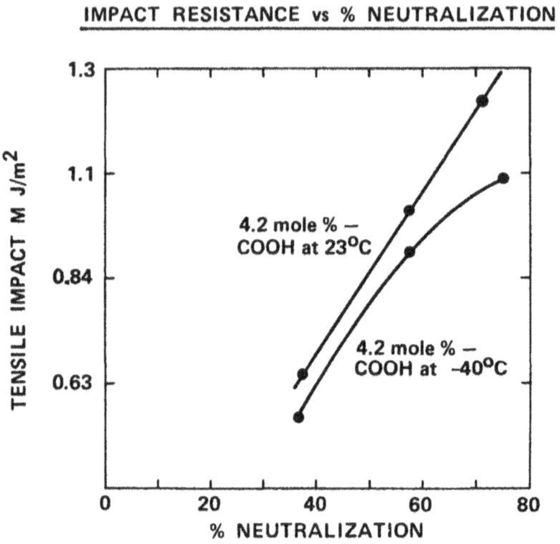

Similarly, the tensile strength, yield strength and hardness of acid copolymers increase as % ionization increases. Furthermore, these properties, plus high melt strength, allow ionomers to be foamed to lower densities and still perform in uses such as bumper guards and softball cores.

RECENT PRODUCT DEVELOPMENTS

Low Modulus Ionomers

Standard ionomers produced from ethylene methacrylic acid are usually rather stiff and hard. However, markets have been identified which require softer, more flexible materials that still retain oil resistance, tensile properties, and abrasion resistance of standard ionomers. To meet these needs, Du Pont has developed copolymers of ethylene, a softening comonomer, and methacrylic acid. The shell and cuff of two-piece ski boots, for example, are made with this low modulus ionomer. Other uses for such resins have been found in footwear and flexible tubing.

The temperature/flexural modulus response of this softer material closely approximates that of a thermoplastic polyurethane containing a polyester elastomeric segment (Graph VIII).

A COMPARISON OF FLEXURAL MODULUS OF LOW MODULUS IONOMER AND COMPETITIVE POLYURETHANE

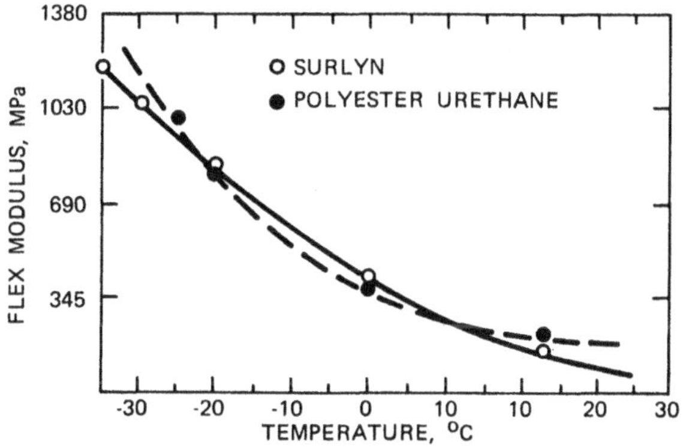

A new low modulus Surlyn® ionomer which is more flexible and has a flatter temperature/modulus response (Graph IX) has been developed. This material closely resembles thermoplastic urethanes based on polyether elastomeric segments. Uses for these resins are being developed in the textile laminate and automotive markets.

High Performance Surlyn®

Although Surlyn® has excellent impact resistance, abrasion resistance and good tensile properties, it has been excluded from certain molding markets normally filled by such materials as polypropylene and acrylonitrile-butadiene-styrene (ABS) copolymers. Surlyn® has been unable to compete with these thermoplastics in uses where high heat deflection temperature and high modulus are required.

Earlier this Spring, Du Pont introduced Surlyn® HP, a family of new ionomer alloys, which

FLEXURAL MODULUS vs TEMPERATURE

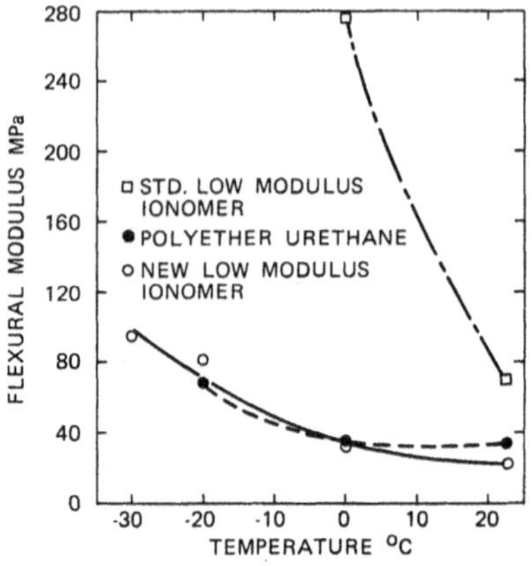

exhibit the toughness properties of ionomers plus stiffness levels that approach those of engineering-type plastics. These resins also have higher resistance to heat than standard ionomers, as demonstrated by higher heat deflection temperature, and they retain the excellent impact strength and abrasion resistance of Surlyn®. In fact, Surlyn® HP's impact strength is equal to or better than that of high impact ABS (Figure VI). Surlyn® HP is being evaluated by automotivbe companies for uses such as air dams and door panels.

COMPARISON OF SURLYN – HIGH PERFORMANCE SURLYN AND ABS

	High Stiffness Ionomer	ABS Extra High Impact	High Performance Surlyn
Density	0.950	1.04	1.08
Hardness	66 (Shore D)	103 (Rockwell)	163 (Rockwell)
Heat Deflection Temp. at 455 KPa °C	43	102	86
Flexural Modulus MPa	380	2280	2800
Tensile Strength MPa	37.2	41.4	25
% Elongation	350	–	25
Notched Izod Impact KJ/m			
at 23°C	0.64	0.35	0.70
at -29°C		0.08	0.4

CONCLUSIONS

This paper discusses the discovery of ethylene copolymer ionomers, and the first attempts at commercialization of these materials as Surlyn®. In addition, the structual features of ionomers and how these features are responsible for the physical properties of ionomers is presented. The importance of these properties in selected packaging and molding end use applications was highlighted. Recent low modulus and high modulus product developments demonstrate that the ionomer technology is one of the most complex and at the same time most interesting fields spawned by polyethylene polymerization technology.

I would like to acknowledge the contributions of many in developing the supporting data and structural studies report here. The bibliography includes many references to the technical literature; but from within the Du Pont Company, I wish to cite particular contributions from Ruskin Longworth, R. J. Powell, R. W. Rees, E. T. Pieski and T. R. Earnest, Jr.

Senior Research Associate; E. I. du Pont de Nemours & Co. (Inc.), Wilmington, Delaware

REFERENCES

1. U.S. 4,351,931 J. B. Armitage
2. Modern Plastics, R. W. Rees, Sept. 1964, Pg.209
3. U.S. 3,264,272 R. W. Rees
4. ACS Polymer Preprints, R. W. Rees and D. J. Vaughan, No. 6, 1965, p. 287, 296
5. Macromolecules, E. P. Otock and T. W. Kewi, No. 1, 1968, p. 401
6. ACS Polymer Preprints, No. 9, 1968, pp. 505-537, 583-617
7. ACS Polymer Preprints, R. Longowrth and D. J. Vaughan, No. 9, 1968, p. 525
8. Macromolecules, E. P. Otocka and D. D. Davis, No. 2, 1969, p. 437
9. Macromolecules, A. Eisenberg, No. 3, 1970, p. 147
10. Macromolecules, T. Kajiyama, T. Oda, R. S. Stein and W. J. MacKnight, No. 4, 1971, p. 198
11. Macromolecules, R. E. Prud'homme and R. S. Stein, No. 4, 1971, p. 668
12. Polymer Letters, C. L. Marx, J. A. Koutsky and S. L. Cooper, No. 9, 1971, p. 167
13. ACS Polymer Preprints, R. J. Roe, No. 12, 1971, p 730
14. ACS Polymer Preprints, A. Eisenberg, No. 14, 1973, pp. 871-896
15. Plastics and Rubber: "Materials and Applications", R. Longworth, Aug, 1978, p 75
16. Journal of Polymer Science: Macromolecular Reviews, W. J. MacKnight, and T. R. Earnest, Jr., Vol. 16, 41-122 (1981)
17. Developments in Ionic Polymers-1, "The Structure and Properties of Ionomers", R. Longworth, Applied Science Publishers, Ltd. Pg. 53

Frederick J. Karol

CATALYSIS AND THE POLYETHYLENE REVOLUTION

INTRODUCTION

The world of polyolefins is enormous (Table I).[1a] Indeed the world of polyethylene itself is very large and is undergoing a major revolution (Table II).[1b-4]. Developments in catalysis for polyethylene in the last decade have had a major impact on the polyethylene industry. After nearly a half century of producing low-density polyethylenes (LDPE) at 20,000-50,000 psi and 300°C (Figure 1), new technology capable of operating at less than 300 psi and near 100°C has emerged. Union Carbide Corporation has developed a unique and versatile low-pressure, gas-phase process that does away with the extremely high pressures and temperatures characteristic of the conventional processes for making low-density polyethylenes. The scientific community and industrial organizations around the world have recognized Union Carbide's gas-phase, UNIPOL process, as a major technological accomplishment.[2] The process couples innovative reactor design and process concepts with special catalyst compositions to achieve a simpler, more versatile, and much more economical route to low-density (LLDPE) and high-density polyethylenes (HDPE) (Figures 2 and 3).

The world of polyethylene is about 50 years old. During this period polyethylene has grown to be the world's largest thermoplastic. High-pressure technology, once dominated by concerns for reactor engineering and reactor control of polymer properties for polyethylene, is rapidly giving way to a new technology in which catalysis and chemical control of product properties are key factors. Key to the success of the UNIPOL process is the proprietary catalysts that operate at low pressure and low temperatures and which are suitable for use in a gas-phase, fluid-bed reactor. It is primarily by means of catalyst composition that resin properties are controlled in the UNIPOL process.

Historians, looking at the changes in the last decade

R. B. Seymour and T. Cheng (eds.), History of Polyolefins, 193–211.

will call this period "Catalysis and The Polyethylene Revolution." This paper discusses the revolution, describes the UNIPOL process, and focuses on the role of catalysis for providing versatile, catalytic polymerization of ethylene in a fluid-bed reactor.

TABLE I
WORLD PLANT CAPACITIES FOR POLYOLEFINS[1a]

YEAR	MM Tons/Yr. LDPE	LLDPE	HDPE	PP
1980	16.10 ⟶		8.48	6.86
1981	16.10	0.44	9.27	7.04
1982	14.48	1.78	8.70	7.27
1983	13.92	2.91	9.40	7.93
1984	14.22	3.92	9.80	8.19
1985	14.62	4.65	10.07	8.81

TOTAL 1984 CAPACITY ∿36 MM TONS (79 MMM LBS)

TABLE II
LDPE AND LLDPE MARKETS IN USA (1980-1985)[1b]

YEAR	MM Tons/Yr. LDPE	LLDPE	% LDPE	% LLDPE
1980	2.59	0.28	90	10
1982	2.32	0.66	78	22
1983	2.40	0.85	74	26
1984	2.35	1.10	68	32
(1985)	(2.30)	(1.41)	(62)	(38)

THE REVOLUTION BEGINS

Discovery by workers of Imperial Chemical Industries, Ltd., of the polymerization of ethylene under high pressure into a solid thermoplastic polymer was an event of immense significance in development of the petroleum chemicals industry.[5-7] From 1933-1938 researchers at ICI were able to define a process for the conversion of ethylene into solid polymers by subjecting it to a pressure of at least 18,000 psi at 150-300°C, long enough to effect polymerization. Based on these early discoveries, an enormous worldwide industry has emerged which has capacity to produce about 14 million metric tons of high-pressure polyethylene.

Discoveries in the laboratories of Ziegler and Natta in the early 1950s caused a revolution in polymer and

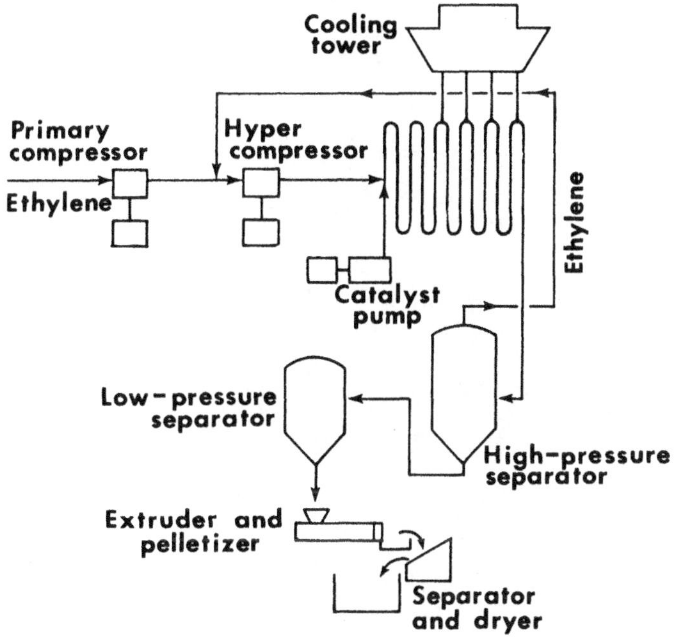

FIGURE 1. HIGH-PRESSURE TUBULAR PROCESS FOR LOW-DENSITY POLYETHYLENE

organometallic chemistry.[8] The ability to polymerize ethylene at atmospheric pressure and room temperature resulted from extensive studies by Ziegler on reactions of organometallic compounds with olefins. In the broadest sense, Ziegler catalysts are reaction products of metal alkyls, aryls, or hydrides of Group I-IV and certain transition metal compounds of Groups IV-VIII. Natta discovered that Ziegler catalysts containing highly ordered transition metal salts in a low valence state (e.g., $TiCl_3$, VCl_3) polymerize α-olefins to crystalline stereoisomeric polymers.

Independent catalyst research, carried out by several U.S. oil companies in the early 1950s with transition metal oxides supported on refractory metal oxides, led to the discovery of some of the earliest low-pressure catalysts for olefin polymerization. These catalysts consist generally of oxides of transition metals from Groups V-VII. For catalytic activity the transition metal oxides are supported on high-surface-area solids such as silica and

FIGURE 2. FLUID-BED PLANT AT SEADRIFT, TEXAS FOR LINEAR, LOW-DENSITY POLYETHYLENE

silica-alumina. A silica-supported chromium oxide (CrO_3/SiO_2) catalyst developed at Phillips Petroleum is the most important transition metal oxide catalyst for high-density polyethylene (HDPE) polymerization.[9] High-density polyethylenes made with the Ziegler-Natta and metal oxide catalysts are predominately linear products with essentially no long chain branching.

DuPont, Canada since about 1960 has operated a solution plant using the SCLAIRTECH process, capable of producing HDPE and LLDPE products. Polymerizations are based on transition metal compounds developed in the DuPont laboratories in the U.S. The original plant was intended to produce 11 thousand metric tons of HDPE/LLDPE resins. A paper [10] presented by DuPont, Canada describes the origins,

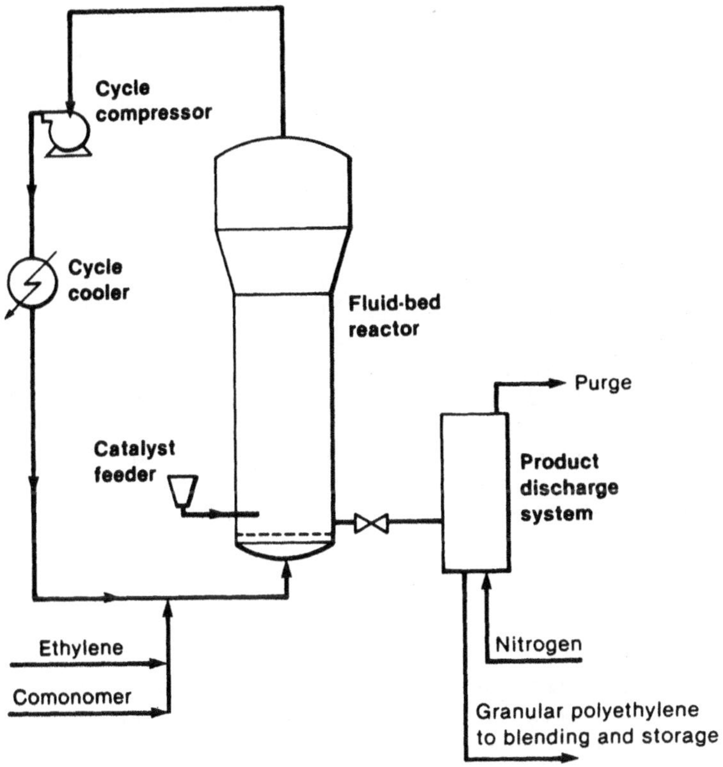

FIGURE 3. LOW-PRESSURE, FLUID-BED PROCESS FOR LINEAR LOW-DENSITY POLYETHYLENE

developments, and features of SCLAIRTECH. Today, DuPont, Dow, and others are actively pursuing the commercial exploration of solution processes for LLDPE. Phillips Petroleum Company announced the development of a low-density slurry process in 1969 and reported production of copolymers at densities down to 0.925 g/cc.[11, 12] However, Phillips elected to concentrate on copolymers of 0.958 down to about 0.93 density, covering the upper half of the linear, low-density range.

Current worldwide capacity for production of HDPE using transition metal catalysts is about 10 million metric tons per year. The HDPE industry, since about 1958, has grown in parallel with low-density polyethylene, both types of polyethylene finding their own end-use applications. The early lead, economics, and production capacity established for high-pressure, low-density polyethylenes did not encourage the development of transition metal catalysts to replace the free-radical, high-pressure route. A host of technical problems in both the process and product areas presented formidable challenges that needed to be overcome. Until the development of the UNIPOL process, no serious attempts were made in the past to replace the high-pressure route with a low-pressure one using transition metal catalysts.

Some activity using Ziegler-Natta catalysts in high-pressure reactors continues.[13] Much of this activity developed in response to the announcements and progress with LLDPE manufactured in low-pressure processes. At high pressure (5,000-30,000 psi), polymerizations with Ziegler-Natta catalysts are carried out at 170-350°C. The low-density products show a broad compositional heterogeneity ascribed to the use of a heterogeneous catalyst. Molecular weight distributions ranged from 3.5-5.7.

KEY ADVANCES LEADING TO THE UNIPOL PROCESS

Research at Union Carbide in the 1960s on low-pressure processes indicated that a gas-phase process had the greatest potential for versatility in producing the full range of molecular weights and densities of polyethylenes. Previous work had shown that polymerizations in a slurry of hydrocarbon diluents limited polymer density to a minimum of about 0.925 to 0.935 g/cc. Excessive swelling of the polymer particle at practical operating temperatures sets this bottom boundary. Solution polymerization requires higher investment costs. Only lower molecular weights could be achieved because of the high solution viscosities encountered with higher molecular weight polymers.

A successful gas-phase process must satisfy several critical requirements which include:

1. Preventing agglomeration of the polymer bed.
2. Control of polymer particle size and shape to achieve effective heat transfer between the resin particles, as well as maintaining proper bed fluidization.
3. Developing highly productive catalysts that efficiently polymerize ethylene and other olefins to yield low-density and high-density products having desired molecular weights and molecular weight distributions.

These requirements were met first in producing medium and high-density polyethylenes during the scale-up of a fluid bed, pilot plant reactor in 1965 and the initial operation of an 8-foot diameter prototype fluid-bed reactor at Seadrift, Texas in 1968.

Further catalyst research was necessary to produce LLDPE structures with commercially attractive properties and under conditions which avoid agglomeration of the polymer bed and reactor fouling. After an aggressive research program, this was accomplished on a pilot scale in the early 1970s. Union Carbide began commercial production of LLDPE in 1975. It announced the UNIPOL Process in November 1977 (Table III).[14]

TABLE III
KEY DATES IN DEVELOPMENT OF UNIPOL PROCESS FOR LLDPE

1950s-1985	Continuous Research with Low-Pressure Catalysts
1965	Pilot Plant for MDPE and HDPE
1968	8-Foot Diameter Prototype Production Reactor
1970-1974	First Generation Catalyst Family for LLDPE
1972-1974	Pilot Plant Production of LLDPE
1975	Commercial Production of LLDPE
1976-1980	Second Generation Catalyst Family for LLDPE
1977	UNIPOL Process Announced
1980-1985	Third Generation Catalyst Family for LLDPE

PROCESS DESCRIPTION AND BENEFITS OF UNIPOL PROCESS

In the UNIPOL process (Figure 2, 3), gaseous ethylene, a comonomer and a catalyst, in the presence of a chain transfer agent, are reacted at low temperatures and pressures to produce a polyethylene product that is removed directly from the reactor as a granular, solid material. Gaseous ethylene, comonomer, and chain transfer agent are fed continuously into the fluidized bed reactor. Catalyst is added separately. Reaction temperature is maintained at less than 100°C, and pressures of only 100-300 psi are required. Circulated by a small compressor, the gas stream fluidizes the polymer bed, provides reactants for polymerization, and removes exothermic heat of reaction. The circulating gas stream passes through a cooler before being returned to the reactor.

Granular polyethylene is removed through a gas lock chamber. Only a small amount of residual monomer accompanies the product into this chamber, and this is safely purged from the resin. Overall the combined conversion rate of ethylene and comonomer is approximately 97% to 99%. The average catalyst residence time is three to five hours during which the polymer particles grow to an average size of 500-1000 microns. The granular product, with or without conventional additives, is ready for packaging and shipping to the customer without any further processing. The size and the shape of the polymer granules are such that their bulk density and solid flow characteristics are suitable for subsequent materials handling and processing operations by the customer. For pelleted product, the granular material may be fed to conventional pelleting equipment or to proprietary Union Carbide pelleting systems.

The UNIPOL process offers significant, commercial benefits (Table IV). Because of the low temperatures and low-pressures used in the UNIPOL process, considerably less space is required for its operation than for high-pressure processes. Most equipment is substantially smaller, and fewer pieces of equipment are needed. For example, a conventional high-pressure process requires an 11,000 HP compressor, which is fed by several smaller compressors totaling several thousand horsepower. The UNIPOL process needs only one 2,000 HP recirculating pump.

No diluents are required in the UNIPOL process. It eliminates the need for elaborate and costly diluent storage, recovery, and purification. The simplicity of the process and the low operating pressures minimize the risk of explosion and considerably reduce the level of hydrocarbon emission. This makes the process acceptable environmentally.

All of these advantages add up to savings of more than 75% in energy consumption and as much as 50% in capital costs compared to a high-pressure facility. Improved technology that will slash operating costs for LLDPE, while simultaneously increasing production capability by 35%, has recently been developed.

TABLE IV
UNIPOL PROCESS FOR LLDPE COMPARED
WITH CONVENTIONAL HIGH-PRESSURE PROCESSES

High-Pressure Process	UNIPOL Process
Operating pressures as high as 50,000 psi	Pressures of less than 300 psi
Temperatures of 300°C	Temperatures of 100°C or less
Long construction lead time	Reduces construction lead time by 8-12 months
Mammoth space requirements	Occupies one-tenth the space of high-pressure plants
Huge capital outlay	Cuts capital costs by as much as 50%
High energy demands	Slashes energy requirements by 75%
Limited to low-density polyethylene	Can produce both high and low-density polyethylenes
Costly and complex maintenance	Easy to maintain
Production rates vary with polyethylene-grade product	Same production rate for all resin grades
Limited turndown capability	Wide turndown latitude
Barely meets environmental requirements	Environmental pollution minimized
Rapidly inflating operating costs	Operating costs reduced
	Wide catalyst flexibility
	Catalyst removal not needed
	Superior resin properties

With the revolutionary developments using low-pressure technology for low-density polyethylenes, a whole new vocabulary for describing this technology has emerged (Table V). Instead of descriptions of free-radical polymerizations involving radical backbiting and long chain branching, the nomenclature now comprises metal catalysis and α-olefin copolymerization. Discussions of autoclave and tubular reactors are giving way to plans for construction of fluid-bed reactors.

PRODUCTS FROM UNIPOL PROCESS

Besides the considerable process advantages for the UNIPOL process, another big payoff comes with the superior properties of LLDPE.[15] Even a cursory comparison of the structures of LDPE and LLDPE suggests that they should show significant property differences (Figure 4). Fortunately these differences translate into better products in numerous end-use applications. The acceptance of LLDPE products in the marketplace has far exceeded most industry forecasts. LLDPE's share of the total U.S. low-density resin market in 1980 was 10%; in 1985 it will grow to 38%, amounting to about 1.4 million metric tons. The largest share is going into film production.

TABLE V

DIFFERENT VOCABULARIES FOR LDPE AND LLDPE TECHNOLOGIES

High-Pressure Process LDPE	Low-Pressure Process LLDPE
Initiators, Free-Radical	Metal Catalysts
Autoclave and Tubular Reactors	Fluid-Bed Reactors
High Temperature	Low Temperature
Radical Backbiting	Copolymerization
Long Chain Branching	Short Chain Branching
Control by Process	Control by Catalyst Composition
High Capital and Energy Demands	Low Demands
Decompositions, Hypercompressors	Granular Polyethylene

The tensile strength of LLDPE is 50 to 75% higher (Figure 5) and elongation at least 50% higher than LDPE products. Other properties such as impact strength and puncture resistance are also greater for LLDPE. At equal density LLDPE has a higher modulus (Figure 6), allowing fabricators to produce lighter or thinner parts with equal stiffness. It also has a broader temperature use range than LDPE. LLDPE resins melt at a 10-15°C higher temperature (Figure 7). The resin also has better low temperature impact strength. In addition to the impressive properties of LLDPE for film applications (Table VI), the LLDPE products show attractive properties for injection molding, rotational and blow molding, pipe and tubing, and wire and cable applications.

TABLE VI
BLOWN FILM MECHANICAL PROPERTY DATA (15)
(C_2/C_4 LLDPE Copolymers vs. LDPE)

		UNIPOL LLDPE		High Pressure LDPE	
Melt Index (gm/10 min)		1.0	2.0	2.0	0.2
Density (gm/cm^3)		0.920	0.920	0.918	0.923
Dart Drop (gms)		145	90	112	185
Puncture Energy (in-lbs/mil)		15.3	14.6	6	4.5
Elmendorf Tear (gm/mil)	MD	120	99	160	90
	TD	340	245	110	100
Tensile Strength (psi)	MD	5880	4970	2900	2800
	TD	4660	3770	2700	3000
Tensile Elongation (%)	MD	620	690	300	300
	TD	760	738	500	500
Tensile Impact Strength	MD	1241	858	440	480
($ft\text{-}lbs/in^3$)	TD	674	433	650	1030
Secant Modulus (psi)	MD	34000	30600	21200	23000
	TD	37500	37000	25700	26000

Conditions: 1.5 mil gauge; 2:1 blow-up ratio

More recent developments have permitted the production of ethylene-higher α-olefin (e.g., hexene-1) copolymers in the UNIPOL process without rate penalties. UNIPOL now has the capability of producing such specialty products that previously were available only from much more expensive processes. The use of higher α-olefins offers the processor products with superior absolute properties, especially those associated with toughness. The processor can also move up to a higher-density grade for greater stiffness while maintaining the same toughness properties as offered by a lower density LLDPE based on butene-1.[16]

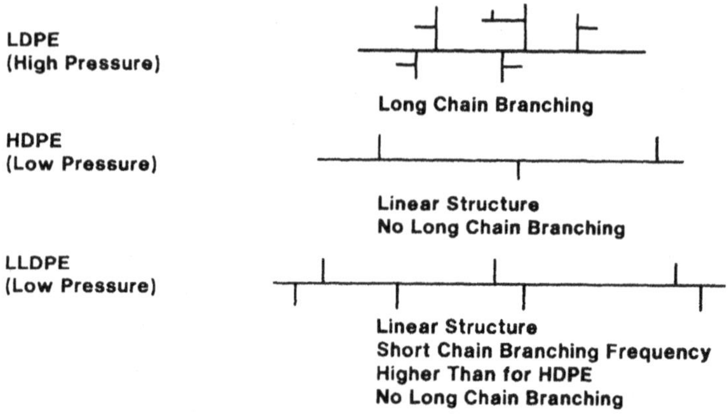

FIGURE 4. STRUCTURES OF CONVENTIONAL LDPE, HDPE, AND LLDPE RESINS

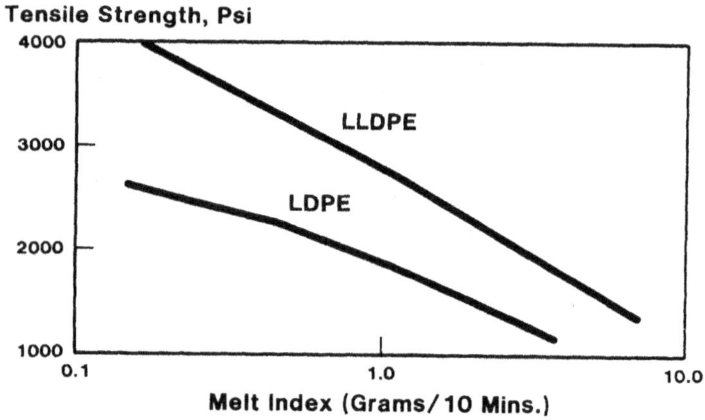

FIGURE 5. TENSILE STRENGTH OF LDPE AND LLDPE

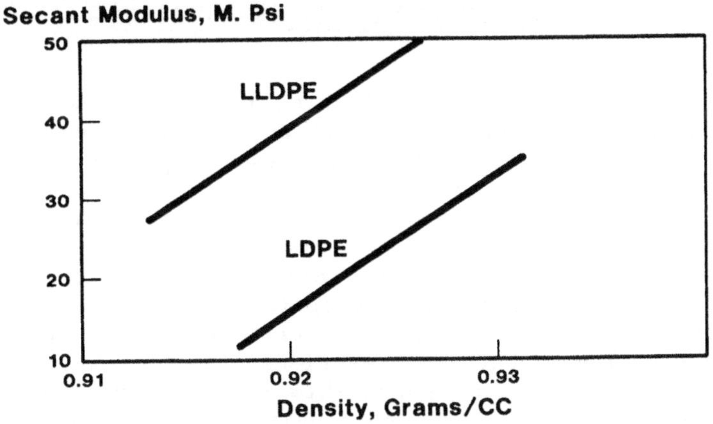

FIGURE 6. MODULUS OF LDPE AND LLDPE

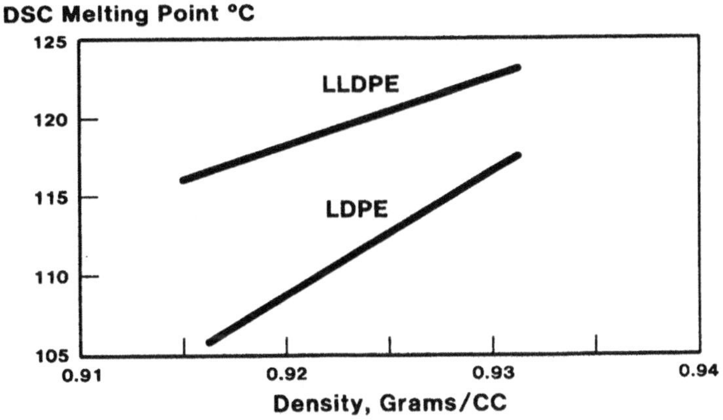

FIGURE 7. MELTING POINT OF LDPE AND LLDPE

CATALYSTS MAKE IT ALL POSSIBLE

Distinctly different reaction processes operate at high pressures to produce LDPE and at low pressure to produce LLDPE. The conventional high-pressure process operates by a free-radical mechanism. Suitable initiators (usually their decomposition fragments) form free-radical species, which add to ethylene to form a primary alkyl radical. Termination occurs via several processes such as disproportionation and radical coupling. Because of the high reactivity of the primary alkyl radicals, branching also occurs. The branching reactions (short chain and long chain branching) are responsible for conferring many of the unique properties to low-density polyethylene. Reactor control of these branching reactions is a key to providing a range of polymer properties.

In the transition metal catalyzed polymerization at low temperatures and pressures, polymerization occurs by a coordinated anionic mechanism:

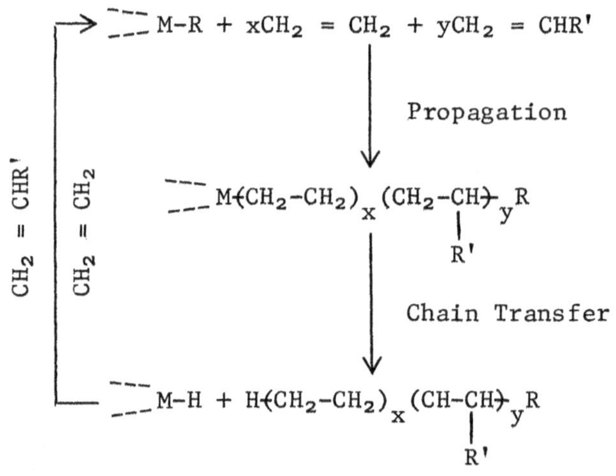

For production of linear low-density products, ethylene is copolymerized with α-olefins such as butene-1 and hexene-1. By combining special coordinated anionic catalysts of high activity and good comonomer response, granular resins over a wide range of densities can be produced. It is the coordination and insertion process that makes the role of the catalyst so significant in influencing process and product behavior. Every molecule that enters the polymer chain must come under the influence of the catalyst and its ligand environment.

Catalysts used in the UNIPOL process satisfy numerous requirements (Table VII). The catalysts should provide such a high level of productivity ($\geq 10^5$-10^6 kg polymer/kg transition metal) that the catalyst residues in the polymer are so small as to allow them to remain in the product and thus avoid the need for catalyst removal steps. High levels of catalyst residues can cause discoloration of the polymer and corrosion of processing machines. Catalyst supports, such as silica or magnesium chloride, to improve polymerization activity by increasing the concentration of active sites, are particularly effective.[17]

Molecular weight of the polyethylenes may be controlled by the reaction temperature and the concentration of chain transfer agent in the system. Hydrogen is an effective chain transfer agent with many catalysts. The specific catalyst type significantly affects molecular weight distribution. Many chromium-based catalysts provide polyethylenes of intermediate or broad molecular weight distribution,[18] while use of titanium-based catalysts lead to polyethylenes of relatively narrow molecular weight distribution.[19-21]

Comonomer incorporation rates are highly dependent on the nature of the specific catalyst used to produce the copolymers. With supported chromium catalysts, ligand environment at the chromium center plays a major role in changing the incorporation rates. The presence of ligands such as the cyclopentadienyl group leads to an increase in the ethylene reactivity ratio. Surface treatment of the support has a small effect on comonomer incorporation. The methods for preparing high activity magnesium-titanium catalysts and the level of electron donor in these catalysts are important parameters in controlling comonomer incorporation rates. A decrease in the electron density at the active titanium center frequently leads to better comonomer incorporation. Improvements in comonomer incorporation lead to higher comonomer efficiency, thereby requiring less dilution of the more reactive ethylene monomer.

Heterogeneous olefin polymerization catalysts can replicate their morphology into the morphology of the polymer particles.[22, 23] The catalyst particles act as a template for growth of the polymer particles. For high activity catalysts in olefin polymerization, the average particle size of the polymer is about 15-20 times larger than the size of the catalyst particle. Although the polymer particle replicates the morphology of the catalyst particle, considerable experimental data show that the catalyst particle breaks down or shatters during the polymerization process and the fragments become dispersed throughout the polymer particle.[24] The polymer initially formed on the catalyst particle may act as a cement for the smaller, shattered particles that bear the active sites for polymerization.

For effective operation and the most efficient use of catalysts in the UNIPOL process, it became necessary to look into heat generation and removal on a microscopic (single particle) scale. By modeling the heat transfer from an individual particle to the gas stream, scientists solved the problem of achieving and maintaining uniformity of heat generation on the microscopic scale. Suitable catalysts were then selected on the basis of uniformity of activity thereby avoiding the higher-than-melting point temperatures that cause polymer particles to agglomerate. With the use of modeling studies and designed experiments, engineers determined the reactor conditions that would assure uniform heat removal and established safe conditions of reaction rate, temperature, gas velocity, and polymer composition.

Final specifications for suitable catalyst compositions used in the UNIPOL process are based on favorable results demonstrated to satisfy all the catalyst requirements. These compositions, defined to provide optimum performance in production and product properties, usually require several years of precommercial development before they can be considered for full-scale commercial use.

TABLE VII
CATALYST REQUIREMENTS IN UNIPOL PROCESS

High Catalyst Productivity
Range of Molecular Weights
Range of Molecular Weight Distributions
Good Comonomer Incorporation
Attractive Particle Morphology
Simple, Reproducible Catalyst Preparation

COMMERCIAL IMPACT AND OUTLOOK FOR THE FUTURE

The impact of this new technology on the plastics industry is awesome. The new low-density polyethylene resins produced by the UNIPOL process were first commercialized in the mid-1970s. Today Union Carbide has programs involving 28 licensees in 15 countries around the world. In 1985 the combined capacity of UNIPOL reactors in operation or under construction around the world will be sufficient to supply 25% of the world's total demand for polyethylene.

It is highly doubtful that the conventional high-pressure process for LDPE would ever have evolved if current catalyst know-how with the UNIPOL process were available in the 1930s and 1940s. The vastly improved economics of the UNIPOL process and the superior properties of LLDPE clearly would have made the low-pressure route the preferred method of manufacture. The current growth in LLDPE reflects the benefits of and opportunities for this fastest growing plastic. Truly, linear low-density polyethylene can be called the industry's wonder child.

The UNIPOL gas-phase process has evolved, first to HDPE, then to low and medium density ethylene-propylene and ethylene-butene copolymers, and more recently to LLDPE containing higher α-olefins. With new technology the process can also now produce unique, very low-density products with densities lower than 0.900 g/cc. These flexible, tough VLDPE products offer significant new product opportunities.[25] In addition, the process is now capable of producing a new line of easy-flow LLDPE resins that combine the excellent strength and drawdown characteristics of general-purpose LLDPE with the easy processibility of a broad molecular weight distribution product.[26]

The UNIPOL process can also produce other polyolefins such as polypropylene, propylene block copolymers, and polybutene-1. A new polypropylene plant of about 80 thousand metric tons per year, based on the UNIPOL process, is going to start production in 1985 in Seadrift, Texas.

The potential of catalyst technology to provide dramatic new improvements in the UNIPOL process and product areas remains high. The 1980s should be a wonderful area of research and development as low-pressure technology and new products from this technology continue to unfold. The excitement over polymerization catalysis, which has already had enormous impact for HDPE and polypropylene production, will continue to encompass the whole polyolefin field.

Frederick J. Karol
Union Carbide Corporation
UNIPOL Systems Department
P.O. Box 670
Bound Brook, NJ 08805

REFERENCES

1a. Digest of Polymer Developments, Springborn Laboratories, Inc., N. Platzer, Ed., Confidential Publ., October 1984.

1b. Plastics Industry Europe, Vol. 8, No. 18, End-September 1984.

2. Chem. Eng., 1979, 86 (December 8), 80-85 (1979 Kirkpatrick Chemical Engineering Achievement Award [Union Carbide Corporation]).

3. F. J. Karol, Chemtech., Vol. 13, (April 1983) pp. 222-228. Reprinted in part with permission of American Chemical Society.

4. R. B. Staub, Paper Presented at Golden Jubilee Conference for Polyethylene 1933-1983, London, June 8-10, 1983, Paper No. B5.4.

5. H. M. Stanley, in Ethylene and Its Industrial Derivatives, S. A. Miller ed., Ernest Benn, Ltd., London 1969, Chapter 1, pp. 28-32.

6. A. H. Willbourn, see reference 4, Paper No. A1.

7. D. W. Ginns, see reference 4, Paper No. A2.

8. J. Boor, Jr., Ziegler-Natta Catalysts and Polymerizations, Academic, New York, 1979.

9. J. P. Hogan, J. Polym. Sci., Part A-1, 8, 2637-2652 (1970); J. P. Hogan and R. L. Banks (Phillips) U.S. Patent 2,825,721 (1958).

10. T. K. Moynihan, see reference 4, Paper No. B5.3.1.
11. Chem. and Eng. News, 47, (17), (1969) 15.
12. J. P. Hogan, see reference 4, Paper No. B3.1.
13. J. P. Machon, in Transition Metal Catalyzed Polymerizations, Part B (R. P. Quirk, ed.), MMI Press Symposium Series Vol. 4, Harwood Academic Publ., New York, 1983, pp. 639-649.
14. Chemical Week, Vol. 121, November 23, 1977, p. 35.
15. W. A. Fraser, L. S. Scarola, and M. Concha, TAPPI Paper Synthetics Course, Cincinnati, Ohio, September 15-17, 1980, pp. 237-248.
16. Plast. World, November 1984, 42 (12), 38-41.
17. F. J. Karol, Catal. Rev.-Sci. Eng., 26 (3 & 4), 557-595 (1984).
18. I. J. Levine and F. J. Karol, U.S. Patent 4,011,382 (1977).
19. G. L. Goeke, B. E. Wagner, and F. J. Karol, U.S. Patent 4,302,565 (1981).
20. F. J. Karol, G. L. Goeke, B. E. Wagner, W. A. Fraser, R. J. Jorgensen, and N. Friis, U.S. Patent 4,302,566 (1981).
21. A. Noshay, F. J. Karol, and R. J. Jorgensen, U.S. Patent 4,482,687 (1984).
22. See reference 8, Chapter 8.
23. J. Wristers, J. Polym. Sci., Polym. Phys. Ed., 11, 1601-1617 (1973).
24. M. P. McDaniel, J. Polym. Sci., Polym. Chem. Ed., 19, 1967-1976 (1981).
25. Plast. World, October 1984, 42 (11), 8-9 and 86.
26. D. E. James, Paper Presented at 1985 BIS Specialty Polyethylene Films Conference, Brussels, Belgium, April 17-18, 1985.

Peter J. T. Tait

THE DEVELOPMENT OF HIGH ACTIVITY CATALYSTS IN α-OLEFIN POLYMERIZATION

The development of high activity catalysts for the polymerization of α-olefins is without doubt one of the greater achievements made in the field of polymer science during the past twenty-five years. In the polymerization of propylene catalyst activities have improved from values of 5-10 kg/gTi to values in excess of 1,000 kg/gTi within a comparatively short period of time. This dramatic increase in catalyst activity has been made possible by a number of contributing discoveries, some of which go back to the early days soon after Ziegler's original discovery.

The history of the development of high activity catalysts for α-olefin polymerization falls conveniently into three periods. During the early period, 1955 and onwards, development was concerned with $TiCl_3$-based catalysts and the so-called first generation of industrial catalysts were produced. During a subsequent period higher activity catalysts based on the modification of $TiCl_3$ with appropriate additivies were formulated and used with considerable success. Finally a third period yielded much more highly active supported-type catalysts, mainly based on the use of $MgCl_2$ and electron donors, although there is here a considerable time overlap.

1. FIRST GENERATION CATALYSTS

Whilst Ziegler's discoveries (1) had highlighted the use of an in situ prepared catalyst, viz.,

R. B. Seymour and T. Cheng (eds.), History of Polyolefins, 213–242.

$$TiCl_4 \downarrow \; \underline{\text{In situ}} \text{ prepared catalyst} \xrightarrow[\text{Inert diluent}]{C_2H_4} \text{Crystalline linear polyethylene}$$

the catalysts used so effectively by Natta (2) in his studies on the polymerization of propylene were based on the use of a preformed crystalline modification of titanium trichloride, prepared by hydrogen reduction, viz.,

$$\underset{\text{Preformed crystalline solid}}{\alpha\text{-}TiCl_3} + AlEt_2Cl \xrightarrow[\text{Hydrocarbon diluent}]{C_3H_6} \text{polypropylene} \xrightarrow{\text{stereoisometric fractionation with boiling n-heptane}} \begin{cases} \text{Crystalline isotactic polypropylene} \\ \text{Atactic amorphous polypropylene} \end{cases}$$

Early progress in catalyst development resulted from the establishment by Natta et al (3) that titanium trichloride could exist in four crystalline modifications.

$$TiCl_4 + H_2 \xrightarrow{400\text{-}800\ ^{\circ}C} \alpha\text{-}TiCl_3$$

$$TiCl_4 + H_2 \xrightarrow[\text{temperatures}]{\text{lower}} \beta\text{-}TiCl_3$$

$$\alpha\text{-}TiCl_3 \xrightarrow{150\ ^{\circ}C} \gamma\text{-}TiCl_3$$

$$\left.\begin{array}{l}\gamma\text{-}TiCl_3 \\ \alpha\text{-}TiCl_3\end{array}\right\} \xrightarrow[\text{grinding}]{\text{prolonged}} \delta\text{-}TiCl_3$$

The α- and γ forms have a lower activity whilst the β-form produces polymer which is essentially amorphous.

Commercial first generation catalysts for the polymerization of propylene, however, own their origins to the investigations of Wilchinsky, Looney and Tornquist (4,5) who discovered that ball milling the product obtained from the reduction of titanium tetrachloride by aluminium metal or aluminium alkyl produced a more active catalyst than pure α-$TiCl_3$ and they concluded from their studies that during the milling process both the α- and γ-forms were converted into the so-called δ-form which exhibited double-layer stacking.

$$TiCl_4 + Al \quad \text{or} \quad TiCl_4 + AlEt_2Cl \quad \text{or} \quad TiCl_4 + Al_2Et_3Cl_3$$

$$\downarrow$$

$$TiCl_n.xAlCl_3$$

Ball mill, Heat treat $\downarrow$ $\qquad 2.5 < n < 3$, $\quad 0 < x < 0.5$

$$\delta\text{-}TiCl_n.xAlCl_3$$

Most first generation catalysts are of this solid solution type, δ-$TiCl_3.0.33\,AlCl_3$, and have been ball milled and heat treated and have surface areas in the range 10-40 m^2/g. Normally very rigorous conditions of preparation are imposed so as to regulate the size and shape of the catalyst particles.

The activity and stereospecificity of these catalysts also depend on the type of aluminium compound used as the activator. Use of triethylaluminium gives a much higher activity but a much lower stereospecificity (6,7) and for this reason the use of diethylaluminium chloride as activator is to be preferred in α-olefin polymerization.

2. MODIFIED TITANIUM TRICHLORIDE CATALYSTS

Early attempts to increase the stereospecificity of $TiCl_3$-based catalysts involved either the addition of Lewis bases together with the activator or preferably prior ball milling a suitable Lewis base with the δ-$TiCl_3.0.33\ AlCl_3$ catalyst. A very wide variety of complexing agents have been used and include ethers, esters, ketones, amines, phosphines, phosphites, organophosphorus compounds, polymeric derivatives etc., and most commercial companies have their own particular recipe for catalyst modification with Lewis bases.

The success of a particular complexing agent depends not only on its intrinsic nature, but also on the mode of addition and on the conditions of concentration and temperature used in the various stages of the catalyst preparation.

The use of complexing agents not only produces catalysts with higher stereospecificities but also can yield catalysts with somewhat higher activities, presumably due to complexation reactions with $AlRCl_2$ compounds formed during the catalyst preparation and which would otherwise impair the catalyst activity.

Mechanisms by means of which electron donors may affect activity and stereospecificity are complex and may include some of the following.

(i) Complex formation of donor (D) with active centres (S):

$$S + D \rightleftharpoons S.D$$

This type of reaction may lead to more hindered centres, or centres of increased stability and chirality, or even blocked centres, depending on the concentration and nature of donor D. It may also include preferential complex formation of donor with more exposed and less stereospecific centres.

(ii) Complex formation of donor with aluminium alkyl (A):

$$A + D \rightleftharpoons A.D$$

This reaction will lead to a reduced alkylation ability and a reduction in the reducing capacity of the aluminium alkyl, and also to a reduction in chain transfer with aluminium alkyl leading to higher molecular weights.

(iii) Complex formation of donor with aluminium alkyl dihalides produced during catalyst formation or other surface alkylation reactions, and leading to increased catalyst activity.
(iv) Catalyst modification due to solvation effects of donor with solid catalyst matrix, e.g. removal of $AlCl_3$ from δ-$TiCl_3.0.33AlCl_3$ - type catalysts.

Some selected examples of Lewis bases used as additives in catalyst preparation are listed in Table 1.

A major advance in the pursuit of catalysts with both increased activity and stereospecificity was made by Solvay & Cie with their discovery of ether treated $TiCl_3$-based catalysts (8.9). The relevant patents describe in an elegant fashion the interesting details of the catalyst preparation which may be formulated as follows:

> $TiCl_4$ in an inert diluent is reduced under mild conditions by $AlEt_2Cl$ at temperatures of 0 ± 2 °C over a period of about 3 h to give the solid product, β-$TiCl_3.x'(AlCl_3).z(AlEtCl_2)$, where $x' = 0.15$ and $z = 0.20$. This product is then treated with diisoamyl ether at 35 °C for 1 h when much of the $AlCl_3$ and $AlEtCl_2$ is dissolved out from the solid matrix. The resulting solid is again isolated and reacted with neat $TiCl_4$ for 2 h at 65 °C to yield a violet product of composition α-$TiCl_3.Al(R_nX_{3-n})_x C_y$, where C is the complexing agent (diisoamyl ether) and $0\geqq n\geqq 2$, $x<0.3$ and $0.11>y>0.009$. The best results are reported when n=1. The resulting high porosity product has a very high surface area in comparison to first generation catalysts usually greater than $150 m^2/g$, and may be used in conjunction with $AlEt_2Cl$ to polymerize propylene and other α-olefins.

Catalysts of this type have a five-fold increase in activity over conventional δ-$TiCl_3$. $0.33AlCl_3$ catalysts and produce only between 2 and 5% amorphous material. Also in slurry type polymerizations these catalysts produce polymer particles with very acceptable particle size and shape characteristics due to the disintegration

Table 1. Some Selected Lewis Bases used as Catalyst Additives

Company	Catalyst System	Patent Number
Montecatini	$TiCl_3/AlEt_3$ + pyridine	U.S. 3 139 418 (1964)
Eastman Kodak	$TiCl_3/AlEt_3$ + HMPT	U.S. 3 186 977 (1965)
Phillips Petroleum Co	$TiCl_3.0.33AlCl_3/AlEt_2Cl$ + aromatic esters	U.S. 3 317 498 (1967)
Mitsui Petrochem Ind	$TiCl_4$ reduced with aluminium alkyl: $TiCl_3$ produced ground with organo-nitrogen or organo-phosphorus compound	Jpn. 79 016 479 (1972)
Kazan Kirov	$TiCl_3/AlEt_2Cl$ + organo-phosphorus compound	U.S.S.R. 434 746 (1972)
Toho Titanium KK	$TiCl_3.0.33AlCl_3$ milled with $(EtO)_4$ Si and methyl methacrylate	Jpn. 50 103 495 (1974)
Plastipolimer Res	$TiCl_3/AlEt_2Cl$ + disopropyl ether	U.S.S.R. 495 326 (1975)
BASF AG	$TiCl_3 0.33AlCl_3$ milled with $Bu_3P/AlEt_2Cl$	DE 2 519 582 (1975)
Shell Int Res Mij BV	β-$TiCl_3$ prepared from $TiCl_4$ + $AlEt_3$ heated to 120 °C with BuCl to give violet $TiCl_3$	Belg. 845 791 (1975)

Table 1 cont.

Company	Catalyst System	Patent Number
Stauffer Chemical Co	$TiCl_3.0.33AlCl_3$ milled with Ph_3PO or benzophenone	DE 2 657 917 (1975)
Mitsubishi Chem. Ind. KK	$TiCl_3$ treated with thioether as used with n-propyl or n-hexyl aluminium alkyl	DE 2 905 330 (1978)
BASF AG	$TiCl_3$ n $AlCl_3$ (n = 0.01-1) ball milled with an ester, $R-O-CO-R_2$ or $R_1-O-CO-CR_2 = CHR_2$ and reacted with an alcohol, e.g., $TiCl_3.0.33 AlCl_3$ and ethyl phenyl acetate ball milled and reacted with n-BuOH	DE 3 021 297 (1980)
Dart Industries Inc	Al reduced $TiCl_3$ ball milled with an electron donor such as R_1-O-R_2 and heat treated so as to remove R_1Cl from the catalyst, e.g., $TiCl_3.33 AlCl_3$ ball milled with anisole	U.S. 4 382 018 (1979)
Northern Petrochem Co	$TiCl_3.33\ AlCl_3$ ball milled with benzoic acid ester, e.g., butyl benzoate	Br. 2 109 800 (1981)

Table 2. Selected Metal Alkyl Reduced Catalysts - Donor Treated.

Company	Catalyst System	Patent No.	Chem.Abstract.
1 Mitsubishi Chem Ind.KK	Solvay-type preparation: I_2 added at the beginning to increase stereospecificity	Jpn. 54 093 691 (1974)	91,212, 024
2 Solvay & Cie	Solvay preparation: final $TiCl_4$ treatment replaced by CCl_4	Fr. 2 320 309 (1975)	87, 202, 358
3 Shell Inst Res Miy. BV	$TiCl_4$ reduced with $AlEt_3$ at $-100\,^oC$ to give β-$TiCl_3$; washed with toluene.	Neth.7 602 920 (1075)	86, 122, 054
4 Exxon Res & Eng	Solvay-type preparation: use of diisopentyl ether.	Br. 1 557 329 (1975)	92, 147, 519
5 Naphthachemie SA	$TiCl_4$ reduced with $AlEt_2Cl$ in the presence of diisopentyl ether.	Ger. 2 807 361 (1977)	89, 164, 197
6 Sumitomo Chem KK	$TiCl_4$ in toluene reduced with $AlEt_2Cl$ in the presence of Bu_2O: treated with I_2 at 100 ^oC	Belg. 104 670 (1977)	90, 104, 670
7 Mitsubishi Chem Ind KK	Reduce $TiCl_4$ with aluminium alkyl treat with polydimethylsiloxane in toluene.	Jpn. 54 079 195 (1977)	91, 124 686
8 Tokuyama SodaCo	Solvay type preparation; $TiCl_4$ reduced with $AlEt_2Cl$; treated with diisopentyl ether then $TiCl_4$ polymerization stage.	Jpn. 54 127 995 (1977)	92, 127, 995

Table 2 cont.

Company	Catalyst system	Patent Number	Chem Abstr.
9 Chisso Corp	Solvay-type preparation: $TiCl_4$ reduced with $AlEt_2Cl$; product treated with diisopentyl ether and $TiCl_4$; treatment with methyl hydrogen polysiloxane.	Jpn. 55 009 657 (1978)	93, 027 093
10 Montedison Spa	$TiCl_4$ reduced with aluminium alkyl in the presence of diatomaceous support; treated with dibutyl ether.	Eur. 5824 (1978)	92 199 063
11 Gulf Oil Corp.	$TiCl_4$ reduced with $Al_2Et_3Cl_3$ + $AlEt_2Cl$; treated with di-n-butyl ether then CO_2.	U.S. 4 330 652 (1980)	-
12 Standard Oil Co. (Ind)	$TiCl_4$ reduced with $AlEt_2Cl$; product treated with propylene; reacted with di-n-butyl ether and di-isoamylether, then $TiCl_4$	BE 890 678	-
13 Chisso Corp.	$TiCl_4$ added to $AlEt_2Cl$ + diisoamyl ether; propylene added, product treated with diisoamyl ether + $TiCl_3$, used with $AlEt_2CO$	BE 893 913 (1982)	-

of the catalyst matrix on interaction with $AlEt_2Cl$.

The preparation involves two highly innovative and important steps, viz., treatment with an ether such as isoamyl ether so as to preferentially extract into solution most of the $AlCl_3$ and to produce a porous and weakly bonded matrix, and transformation of the β-$TiCl_3$ into what now appears to be mainly the α-modification by treatment with neat $TiCl_4$ at 65 °C. This latter reaction being much used in subsequent catalyst preparations.

Other industrial groups have produced somewhat similar catalyst systems and a limited selection is listed in Table 2.

3. Highly Active Magnesium Chloride Supported Catalysts.

3.1 Catalyst Systems

Throughout the history of α-olefin catalysis the search for ever better industrial catalysts has been a constant driving force and has produced many of the interesting scientific discoveries which have been made. The optimal characteristics of a good industrial catalyst may be described as follows (10).

(i) A very high level of activity. There is a real economic advantage in using catalysts with such high activities that both the need for catalyst deactivation and the removal of catalyst residues is eliminated. In particular residual levels of chlorine should be less than 50 ppm and levels of titanium less than 10 ppm.

Neither of these criteria were realized in the use of $TiCl_3$-based catalysts which have been described so far in this paper.

(ii) A very high yield of stereoregular high molecular weight polymers.

Elimination of expensive operations to remove atactic and low molecular weight fractions has become necessary because of energy considerations.

(iii) Good copolymerization behaviour.

The development of high activity catalysts having the ability to copolymerize a range of comonomers is of high commercial interest, and this is an area of polymerization catalysis which is still not well understood.

(iv) Spherical particle shape and narrow particle size distribution.

It has now been clearly demonstrated that the size and shape of the catalyst particles control the morphological properties of the resulting polymer particles (11). The production of polymers with controlled particle size and bulk density has become a very important consideration in the design of acceptable catalysts, and this is particularly true for gas phase polymerizations.

For a catalyst to perform in a satisfactory manner it is necessary for the catalyst granules to have a high surface area, to be highly porous, to be fracture resistant (and hence eliminate the production of fragmented angular particles), and to have a uniform distribution of active centres.

(v) Selective control of molecular weight and molecular weight distribution.

The sensitivity of polymerization centres towards temperature and chain transfer agents such as hydrogen is of paramount importance in catalyst behaviour.

The development of more active catalysts by means of either Lewis base addition or ether treatment as adopted by Solvay Cie, whilst leading to improved catalysts, only partially fulfilled the necessary requirements as have been detailed (10). More fruitful developments came from another parallel course of catalyst development which had its origins in the discovery by Hogan and Banks (12,13) in 1958 that chromium trioxide supported on silica (CrO_3/SiO_2) could polymerize ethylene under comparatively mild conditions to produce predominantly linear, high density polyethylene. This discovery together with the determinations of active centre concentration conducted by Natta (14) which demonstrated that only a small percentage of the titanium atoms (< 1%) were active in polymerization indicated the direction in which further developments would take place.

Early experimentation, during 1960-63, involving supporting matrixes such as SiO_2, Al_2O_3, SiO_2, Al_2O_3, ZrO_2, TiO_2, MgO, etc (15) was successful in producing catalyst systems active for the polymerization of ethylene, but when these catalyst systems of this type were used for propylene polymerization they exhibited very poor stereospecificity. A similar situation was found to be true for a wide range of catalysts derived from magnesium compounds, viz; $Mg(OH)_2 + TiCl_4$; $Mg(OH)Cl + TiCl_4$; $MgSO_4.3\,Mg(OH)_2 + TiCl_4$; $Mg(OR)_2 + TiCl_4$; $MgCl_2 + TiCl_4$; $MgCl_2.nD + TiCl_4$ (D≡electron donor compound). Some of these systems, however, have been successfully used for the polymerization of ethylene.

Of greatest significance has been the discovery in 1968 by the Montecatini Edison Co (16) and the Mitsui Petrochemicals Ind (17) that catalysts prepared from magnesium chloride, titanium tetrachloride and an electron donor, and activated by a mixture of trialkylaluminium and an electron donor could polymerize propylene with a high yield (> 50 kg polypropylene/gTi.h) and with high stereospecificity (isotactic index > 90%). The initial patents describe two basic routes for the preparation of these highly active catalysts.

Typically a 30-fold molar excess of very carefully dried anhydrous $MgCl_2$ is ball-milled with a $TiCl_4$-electron donor complex, such as $TiCl_4$.ethyl benzoate, washed with n-heptane, and dried. The catalyst is then activated by a mixture of trialkylaluminium (TAA) and ethyl benzoate (EB) such that TAA:EB:Ti ≈ 300 : 100 : 1.

Alternatively, dried anhydrous $MgCl_2$ is ball-milled for 20 h at 0-5 °C with EB ($MgCl_2$: EB ≈ 1 : 0.15) and then treated with neat $TiCl_4$ at 80 - 130 °C for 2 h, washed with n-heptane, and dried, yielding a pale yellow solid catalyst containing typically 1-5% Ti and 5-20% EB. This catalyst, now sometimes called the procatalyst, is then activated by treatment with a mixture of trialkylaluminium and an electron donor (e.g. ethyl benzoate, p-ethyl anisate, p-methyl toluates, etc.)

The subsequent period since 1968 has been one of intense activity with the publication of many patents which claim high activity catalysts for propylene and other α-olefin polymerization. A selected list of examples is listed in Table 3 and further examples are to be found in the literature (18,19,20).

The use of magnesium chloride in an 'activated' form has been known for some time, and was mentioned by Bryce-Smith in 1959(21) and described by Kamienski (22) in 1967. 'Activated' magnesium chloride may be prepared by a number of methods which include; (i) treatment of $MgCl_2$ with activating agents such as electron donors; (ii) ball-milling $MgCl_2$; and (iii) reaction of compounds such as Grignard reagents with chlorinating agents. As far as use in catalyst preparation is concerned three procedures for activation have been found to be of particular advantage, viz:

i) Ball milling magnesium chloride with an electron donor.
ii) Precipitation of a magnesium chloride compound or complex from solution in such a way as to give catalyst particles of controlled size and shape.
iii) Reaction of Grignard reagents or magnesium alkyls with chlorinating agents either in the presence or absence of a donor.

3.1.1 Effects of ball milling magnesium chloride with electron donors (ethyl benzoate). Magnesium chloride has been found to be uniquely suitable as a support in propylene polymerization. The nearest substitute being manganese dichloride. The reason for this unique behaviour is probably the similarity between its crystal structure and that of titanium trichloride. Both substances have layer lattice structures with close packed atoms, and additionally the ionic radii of Ti^{4+} and Mg^{2+} are similar, being 0.68 and 0.65 Å respectively.

Magnesium dichloride normally exists as a

Table 3. Some Selected Magnesium Chloride Catalysts

Company	Catalyst System	Patent Number
Phillips Petroleum	$MgCl_2$ milled with $TiCl_4$ ethyl benzoate complex and durene; used with $AlEt_3$ + ethyl anisate.	Belg. 843 224 (1975)
Montedison Spa	$MgCl_2$ milled with $TiCl_4$ ethyl benzoate complex; used with $AlEt_3$ + methyl p-toluate	Belg. 845 593 (1975)
Mitsui Petrochem. Ind. KK	$Mg(OPh)_2$ milled with ethyl benzoate; reacted with $TiCl_4$; used with $AlEt_3$ and p-toluate	Belg. 856 189 (1976)
Mitsui Toatsu Chem. Inc.	$MgCl_2$ milled with $AlCl_3$ ethyl benzoate; treated with $TiCl_4$; used with $AlEt_3$	DE 2 734 652 (1976)

Table 3. continued....

Company	Catalyst System	Patent Number
Mitsui Petrochem. Ind. KK	$MgCl_2$ milled with ethyl benzoate and silicone oil; p-cresol added then $AlEt_3$; then $TiCl_4$; used with $AlEt_3$ + methyl p-toluate	DE 2 809 318 (1977)
Exxon Res. and Eng. Co.	$MgCl_2$ milled with $TiCl_4$ ethyl benzoate complex; product treated with $TiCl_4$; used with $(Me_3SiCH_2)_2$ AlEt	Eur. 4 739 (1978)
BASF AG	$Mg(OEt)_2$ milled with PhCOCl; treated twice with $TiCl_4$; used with Et_3Al + ethyl anisate	DE 2 831 830 (1978)
Montedison Spa	MgO treated with $SOCl_2$; product treated with ethyl benzoate and EtOH and then $TiCl_4$	Belg. 875 494 (1978)

cubic close-packed structure of double chlorine layers with interstitial Mg^{2+} ions in 6-fold coordination (23). This layer structure is of the $CdCl_2$ type and leads to a characteristic X-ray diffraction spectrum with a strong reflection (104) at d = 2.56 Å (24). Another less stable crystalline structure is also known (25), and has a hexagonal close packed structure similar to that of α-$TiCl_3$ with a strong reflection (104) at d = 2.78 Å.

During the ball milling process it is observed that there is a gradual disappearance of the (104) reflection and the appearance of a broad halo indicating stacking faults (24). The stacking sequence no longer corresponds to that of a cubic close-packed structure. The spacing of the broadened halo (d = 2.65 Å) lies between those observed for the cubic and hexagonal structures. Galli et al (26) have concluded that the best model describing the structural disorder of activated $MgCl_2$ (in the presence of $TiCl_4$) is similar to that proposed by Allegra and Bassi (27) for δ-$TiCl_3$.

During ball milling of $MgCl_2$ in the presence of ethyl benzoate the aggolmerates of small primary crystals are broken down, and the surfaces coated with ethyl benzoate. At the same time it is likely that ethyl benzoate will be absorbed between layers in the $MgCl_2$ structure, or that solid solution formation takes place. The ethyl benzoate thus acts the dual role of preventing reaggregation of primary crystallites (c.f., protected colloids)(28), and of penetrating the $MgCl_2$ matrix, at least in the surface vicinity.

It is widely believed that the ethyl benzoate is coordinated to the $MgCl_2$ via the C = O group. Goodall (29) has reported that the C = O stretching frequency changes from 1721 cm in the free ester to 1680 cm^{-1} in the adsorbed ester, and similar results have been observed by Kashiwa(30) and Chien et al(31).

3.1.2 Treatment of ball milled $MgCl_2$ - electron donor (ethyl benzoate) with titanium tetrachloride. There is good experimental evidence which suggests a strong interaction between $TiCl_4$ and the $MgCl_2$ - ethyl benzoate support. The second step in the catalyst preparation involves treatment of the solid products from ball milling with hot neat $TiCl_4$ at 80-130 °C, or by ball milling the solid products with $TiCl_4$. During treatment with $TiCl_4$ ethyl benzoate is extracted by the $TiCl_4$ and a typical catalyst will only contain some 5-20% ethyl benzoate after treatment. Ethyl benzoate is lost not only from weakly bonded sites on the lateral faces (1700 cm^{-1} absorption) but also from the more strongly coordinating corner sites (1650 cm^{-1} absorption) (29), and it is believed that the $TiCl_4$ coordinates to those sites by replacement of the ethyl benzoate molecules. Whilst this sort of replacement reaction certainly takes place it is more likely that the $TiCl_4$ is adsorbed as a complex with ethyl benzoate; the effect of ethyl benzoate on the stereospecificity of the polymerization providing strong evidence for this suggestion. This reaction may be regarded as formation of a solid solution of the complex and the catalyst matrix as with $AlCl_3$ and $TiCl_3$ (4,5). One further observation which supports a strong interaction between the $TiCl_4$ and the $MgCl_2$ - ethyl benzoate support is that it is impossible to completely extract all the $TiCl_4$ from the $MgCl_2$ after the catalyst preparation (11).

3.1.3 The cocatalyst system. In order to improve the stereospecificity of the polymerization an electron donor is normally used along with the trialkylaluminium compound which is used as activator (16,17). Benzoates, toluates, anisates and various amines have been used extensively with $MgCl_2$/ ethyl benzoate/ $TiCl_4$ catalyst systems while various di- and trialkoxysilanes have been used effectively with $MgCl_2$/phthalate/$TiCl_4$ systems. The role of an electron donor when used in this manner may be detailed as follows:

(i) Complex formation with aluminium alkyl

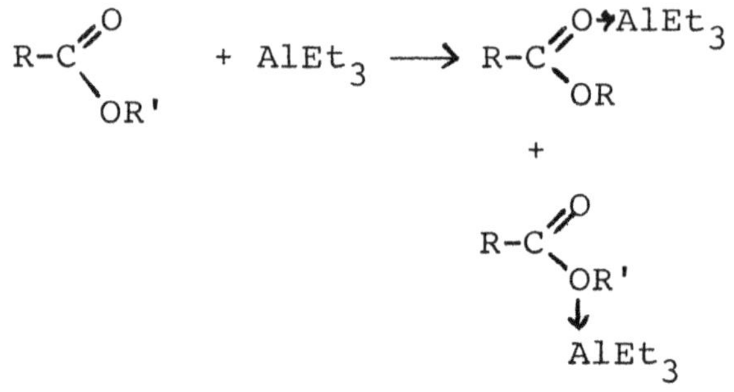

This reaction will reduce the alkylating ability of the aluminium alkyl and also decrease its capacity to reduce the supported titanium entity. Additionally complex formation between trialkylaluminium and an electron donor will decrease chain transfer leading to increased molecular weights (32).

(ii) Increased stereospecificity.
The addition of a suitable donor to the activator normally leads to a dramatic increase in stereospecificity which is believed to arise from either conversion of aspecific centres into isospecific centres, or deactivation of the aspecific centres by selective complex formation, or a combination of both processes. Evidence to-date would seem to support the hypothesis that both these processes may take place, depending on the relative concentrations of donor, trialkylaluminium and active centres, as well as the order of mixing (33).

(iii) Alkylation and reduction of donor.

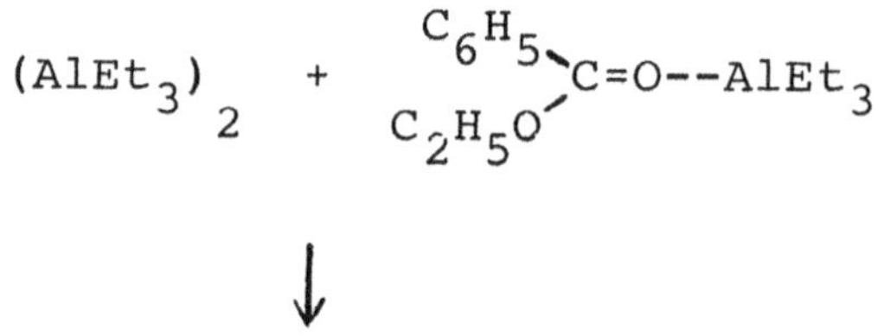

$$\mathrm{Et_2AlOC_2H_5} \quad + \quad \mathrm{Et_2AlOC{-}}\underset{\underset{\mathrm{C_2H_5}}{|}}{\overset{\overset{\mathrm{C_2H_5}}{|}}{\mathrm{C}}}\mathrm{{}_6H_5}$$

Alkylation, reduction and eventual elimination reactions between trialkylaluminium and donor have been shown to take place (34-37), and the products formed may themselves act as catalyst modifiers. Indeed the reactivity of esters towards $AlEt_3$ is dependent strongly not only on the molar ratio of the components and on the temperature, but also on the concentrations and solvent. Goodall (29) has reported on the interaction of p-ethyl anisate and $AlEt_3$, while Chien et al (38) have investigated the Interactions between methyl p-toluate and $AlET_3$ and ethyl benzoate and $AlEt_3$. More recently Kissin and Sivak (39,40) have carried out a detailed investigation of the mechanism of catalyst modification in $MgCl_2$/aromatic ester/$TiCl_4$-$AlEt_3$ systems and have suggested that mixtures of aluminium alkoxides are probably the true modifiers of such supported catalysts for olefin polymerization.

The exact role of electron donors when added to the cocatalyst is still unresolved but their presence is a crucial feature of these polymerizations, affecting both catalyst activity and stereospecificity. Where chemical reaction between electron donor and aluminium alkyl readily takes place the use of much larger concentrations of aluminium alkyl becomes necessary, and for this reason catalyst preparations, either avoiding the use of such donors, or using modifiers already built into the procatalyst system itself, offer considerable advantages.

3.2 Control of Shape of Catalyst Particles.

The relationship between the shape of the catalyst particles and the morphological properties of the resulting polymer has been realized for many years and featured in many of the earlier preparations of $TiCl_3$-based catalysts, e.g., catalysts prepared from $TiCl_4$ and $Al_2Et_3Cl_3$ and used in a modified form for the polymerization of propylene. The replication of catalyst particle shape in the polymer granules has been clearly established by Galli et al (11,41) for catalyst systems based on magnesium chloride as is shown in Fig. 1.

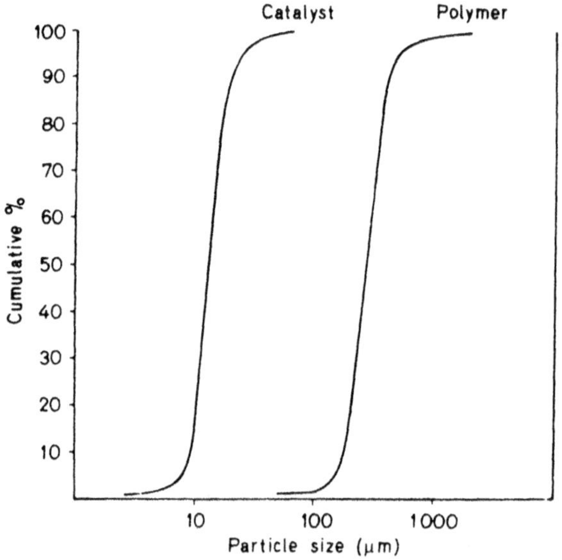

Fig. 1 Particle size distributions (replica).

As a result of the economic implications of avoiding the granulation step in polymer production it has become desirable to produce spherical catalyst particles with a controlled and narrow molecular weight distribution.

Considerable effort has been made to prepare magnesium chloride particles in this form, and an interesting method has been developed by the Mitsui Petrochemical Industries Ltd. (42) whereby such a catalyst is prepared via the use of the alcoholate, $MgCl_2.xEtOH$.

4. Kinetic Behaviour of High Activity Magnesium Chloride Supported Catalysts.

4.1 Rate-time Behaviour

Catalysts of the type $MgCl_2$/donor/$TiCl_4$-$AlEt_3$/donor when used for the polymerization of propylene are characterized by having very high activities and also high stereospecificities. Most catalyst systems are also characterized by having exceedingly high initial rates of polymerization which decrease rapidly with time, as is shown in Fig. 2.

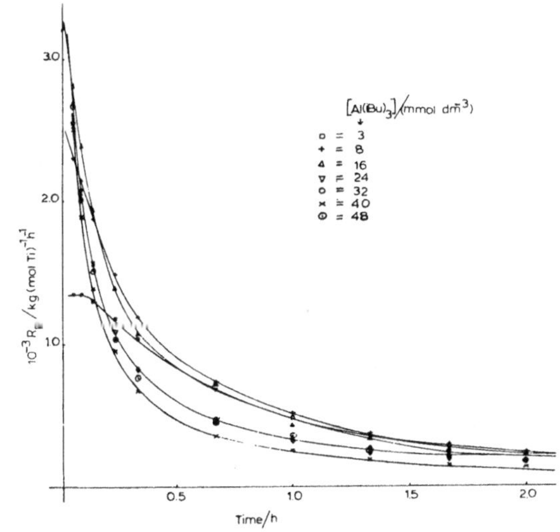

Fig. 2 R_p versus time plots for propylene polymerization using $MgCl_2$/ethyl benzoate/$TiCl_4$-$Al(i\text{-}Bu)_3$ catalyst system at 60 °C.

Whilst many of these rate-time decay-type plots can be fitted to second order decay plots the situation is more complex and has been discussed elsewhere (28,43,44).

Other catalyst systems, viz., many phthalate-based catalyst systems, behave in a very different manner and show very constant rate-time plots, as is shown in Fig. 3.

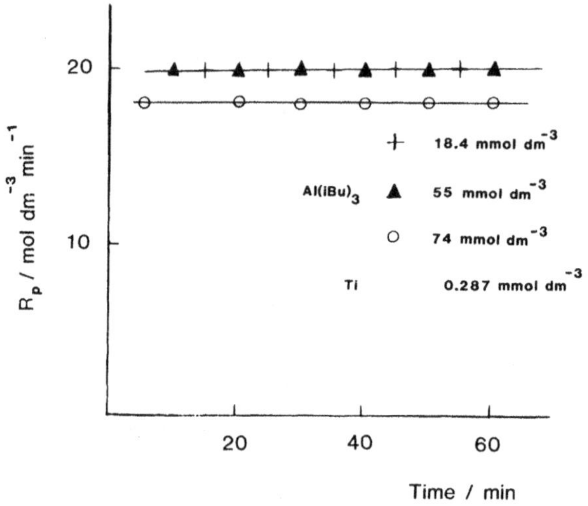

Fig. 3 R_p versus time plots for 4-methylpentene-1 polymerization using $MgCl_2$/phthalate/$TiCl_4$-$Al(C_6H_{13})_3$ catalyst system at 40 °C.

4.2 Active Centre Concentration and Propagation Rate Constants

Due to the higher activity shown by $MgCl_2$-supported catalysts for α-olefin polymerization there has been considerable interest in the determination of active centre concentration, C*, and in the evaluation of values for rate constants for chain propagation, k_p. Progress in these areas has been reviewed recently

(28,45), and in recent publications Tait and Abu Eid (46) and Chien et al (47) provide further detailed results and analyses. Whilst it is apparent that different numerical values for C* and k_p are obtained for catalysts prepared using different routes and using differing conditions of preparation, certain conclusions emerge from these studies.

(i) Higher active centre concentration values. Whilst not all the Ti atoms in $MgCl_2$/donor/$TiCl_4$ - $AlEt_3$/donor catalyst systems are active in the polymerization of α-olefins, values for C* between 2-60% have been reported (28,46,47). Some typical values are listed in Table 4. These values are much higher than experienced normally in $TiCl_3$-based catalysts. It must also be realized that because of morphological considerations governing the growth of polymer chains throughout the mass of the catalyst granules that not all the active centres may be active at the same time during polymerization (11) and further results on this situation will be presented in a later publication (52). However the same situation may arise because of the fundamental nature of the mechanism of Ziegler-Natta polymerization (53). Furthermore it must be realized that not all active centres will be of the same kind or even show the same activity (47,54).

(ii) Higher values for the propagation rate constant.

Higher values for k_p have also been reported, especially during the initial stages of the polymerization (48-51), and some typical values are listed also in Table 4.

Some more recent results for C* and k_p values obtained for the polymerization of 4-methylpentene-1 are presented in Table 5 which also lists values for C* and k_p obtained using titanium and vanadium-based catalysts under comparable conditions. It is evident that much of the increased activity shown by $MgCl_2$-supported catalyst systems when used for the polymerization of 4-methylpentene-1 arises from

Table 4. Selected values of C* and k_p for various $MgCl_2$ supported catalysts.

	Catalyst System	Monomer	Time	C*/ mol%Ti	k_p/ $dm^3mol^{-1}s^{-1}$	Ref.
1.	$MgCl_2$ supported high activity catalyst	propylene	5min 30min 180min	8-10*	500 300 140	(48)
2.	$MgCl_2/EB/TiCl_4-AlEt_3$ (Ti content = 2%Ti)	propylene	7-120s	2.8*	2700	(49)
3.	$MgCl_2/TiCl_4 \cdot EB-AlEt_3$ (Ti content = 2.5%)	propylene		0.27* 0.68*	870* 1250*	(51)
4.	$MgCl_2/TiCl_4-AlEt_3$	propylene	5s	20-60 2-6*	240- 730 500-1500	(50)
5.	$MgCl_2$ supported high activity catalyst	but-1-ene	300s		53	(48)

* isotactic fraction

Table 5. Comparative values of C* and k_p for selected catalyst systems for 4-methylpentene-1 polyermerization.

Catalyst system	Solvent	C*/ mol % Ti	10^{-3} k_p/ min^{-1} *
$Mg_2Cl_2/EB/TiCl_4$ - $Al(i\text{-}Bu)_3$	Heptane	29	4.67
Stauffer 1.13 - $Al(i\text{-}Bu)_3$	Heptane	0.68	1.29
VCl_3 - $Al(i\text{-}Bu)_3$	Benzene	0.038	3.10

* Calculated using $R_p = k_p \theta_M C^*$

the considerable increase in C* (X43) and also from a more modest increase in k_p.

The electronic environment of Ti atoms located on or in $MgCl_2$-type matrixes will differ from that of Ti atoms located on or in $TiCl_3$-type matrixes and so some difference in relative k_p values would be expected.

Any study of the history of the development of catalysts for the polymerization of α-olefins is in many ways exciting and rewarding. The response, in many cases, to technological demands, has yielded fresh and important discoveries which have been pioneered often in industrial laboratories throughout the world. Nevertheless, we must still await with expectation a better and fuller fundamental understanding of many of the chemical and physico-chemical reactions involved in catalyst formation and in these complex polymerization systems. The past has been a harvest rich in endeavour and in concept and we await the future with eager expectation.

Department of Chemistry,
UMIST,
Manchester M60 1QD,
UNITED KINGDOM

REFERENCES

1. K. Ziegler, E. Holzkamp, H. Breil, and H. Martin, Angew. Chem., 67, 541 (1955)
2. G. Natta, J. Polym. Sci., 16, 143 (1955)
3. G. Natta, P. Corradini, and G. Allegra, J. Polym. Sci., 51, 399 (1961)
4. Z.W. Wilchinsky, R.W. Looney, and E.G.M. Tornqvist, J. Catal., 28, 352, (1973)
5. E.G.M. Tornqvist and A.W. Langer, Jr., U.S. Pat., 3 032 510, filed June 27th, 1958, issued May 1st, 1962.

6. G. Natta (Montecatini), U.S. Pat., 3 197 452 (1965)

7. A.A. Korotkov and Li. Tszum - Chan., Polym. USSR (Engl. Transl.), 3, 621 (1962)

8. J.P. Hermans and P. Henrioulle (Solvay & Cie), U.S. Pat., 3 769 233 (1973)

9. J.P. Hermans and P. Henrioulle (Solvay & Cie), U.S. Pat., 4 210 738 (1980)

10. P.J.T. Tait in 'Developments in Polymerization, Vol. 2, ed. R.N. Haward, Applied Sci. Publ., London, p.81 (1979)

11. P. Galli, L. Luciani and G. Cecchin, Angew. Makromol. Chem., 94, 63 (1981)

12. J.P. Hogan and R.L. Banks (Phillips Petroleum Co.), U.S. Pat., 2 825 721 (1958)

13. H.R. Sailors and J.P. Hogan, Macromol. Sci. Chem., A7, 1377 (1981)

14. G. Natta, J. Polym. Sci., 34, 21 (1959)

15. Br. Pat., 969 761; 969 767; 916 132; 1 038 882

16. Montecatini Edison Co., Br. Pat., 1 286 867 (1968)

17. Mitsui Petrochemical Ind., Ital. Pat., 912 345 (1968)

18. T.W. Taylor, K.Y. Choi, H. Yvan and W.H. Ray in 'Transition Metal Catalysed Polymerizations, Alkenes and Dienes, Part A', ed. R.P. Quirk et al, Harwood Academic Press, New York, 191 (1983)

19. P.D. Gavens, M. Bottrill, J.W. Kelland and J. McMeeking in 'Comprehensive Organometallic Chemistry', ed. G. Wilkinson, F.G.A. Stone and E.W. Abel, Pergamon Press, Oxford, Vol. 3, 22.5, p.475 (1982)

20. C. Dumas and C.C. Hsu, Rev. Macromol. Chem. Phys., C24(3), 355 (1984)

21. G.D.R. Pat., 955 807 (1959)

22. C.W. Kamienski, 'Synthesis and properties of diorganomagnesium compounds', Ph.D. thesis, University of Tennessee (1967)

23. G. Bruni and A. Ferrari, Rend. Accad. Naz. Lincei, 2, 457 (1925)

24. U. Giannini, Makromol. Chem. Suppl., 5, 216 (1981)

25. I.W. Bassi (unpublished results, c.f. ref. 24)

26. P. Galli, P. Barbe, G. Guidetti, R. Zannetti, A. Martorana, A. Marigo, M. Bergozza and A. Fichera, Eur. Polym. J., 19, 19 (1983)

27. G. Allegra and I.W. Bassi, Gazz. Chim. Ital, 110, 437 (1980)

28. P.J.T. Tait, Spec. Periodic Rep., Macromol. Chem., 3, 3 (1984), Royal Soc. Chem., London

29. B.L. Goodall in 'Transition Metal Catalysed Polymerizations, Alkenes and Dienes, Part A', ed. R.P. Quirk et al., Harwood Academic Press, New York, p.355 (1983)

30. N. Kashiwa in 'Transition Metal Catalysed Polymerizations, Alkenes and Dienes, Part A', ed. R.P. Quirk et al, Harwood Academic Press, New York, p.379 (1983)

31. J.C.W. Chien, J-C. Wu and C.-I. Kuo, J. Polym. Sci., Polym. Chem. Ed., 20, 2019 (1982)

32. P.J.T. Tait and M. Abu-Eid (to be published)

33. P.J.T. Tait and A.E. Enenmo (to be published)

34. S. Pasynkiewicz and K.B. Starowieyski, Rocz. Chem., 41, 1139 (1967)

35. S. Pasynkiewicz, L. Kozerski and B. Grabowski J. Organometal. Chem., 8, 233 (1967)

36. K.B. Starowieyski, J. Organometal. Chem., 117, Cl (1976)

37. Y. Baba, Bull. Chem. Soc. Jpn., 41, 1022 (1968)

38. J.C.W. Chien and J.-C. Wu, J. Polym. Sci., Polym. Chem. Ed., 20, 2445 (1982)

39. A.J. Sivak and Y.V. Kissin, J. Polym. Sci., Polym. Chem. Ed., 22, 3739 (1984)

40. Y.V. Kissin and A.J. Sivak, J. Polym. Sci., Polym. Chem. Ed., 22, 3747 (1984)

41. P. Galli, P.C. Barbe and L. Noristi, Angew. Makromol. Chem., 120, 73 (1984)

42. Mitsui Petrochem. Ind. Ltd., Australian Patent, 894 621 82

43. T. Keii, Makromol. Chem., 183, 2285 (1982)

44. S. Wang and P.J.T. Tait (in the course of publication)

45. P.J.T. Tait in 'Transition Metal Catalysed Polmerizations, Alkenes and Dienes, Part A', ed. R.P. Quirk et al, Harwood Academic Press, New York, p.115 (1983)

46. P.J.T. Tait and M. Abu Eid, 'Proceedings of the 9th Iberoamerican Sym. on Catalysis' Vol. 1, p.163, Lisbon (1984)

47. J.C.W. Chien, C.-I. Kuo and T. Ang, J. Polym. Sci., Polym. Sci. Ed., 23, 723 (1985)

48. U. Giannini, Makromol. Chem. Suppl., 5, 216 (1981)

49. N. Kashiwa and J. Yoshitake, Makromol. Chem. Rapid Commun., 3, 211 (1982)

50. N. Kashiwa and J. Yoshitake, Makromol. Chem. Rapid Commun. 4, 41 (1983)

51. G.D. Bukatov, S.J. Shepelev, V.A. Zakharov, S.A. Sergeev and Y.I. Yermakov, Makromol. Chem., 183, 2657 (1982)

52. P.J.T. Tait, I.D. McKenzie and D.W. McLelland (to be published)

53. P.J.T. Tait and I.D. McKenzie, Polymer, 13, 303 (1972)

54. P.J.T. Tait in 'Preparation and Properties of Stereoregular Polymers', ed., R.W. Lenz and F. Cardelli, D. Reidel, Dordrecht, Holland, p.85 (1980)

Kazuo Soga

THE ROLE OF $MgCl_2$ ON THE ACTIVITY OF OLEFIN POLYMERIZATION

Polymerizarion of propylene was conducted over the well-defined $MgCl_2/TiCl_4/SiO_2/AlEt_3$ catalyst by changing the molar ratio of Mg/Ti. The addition of a small amount of $MgCl_2$ ($Mg/Ti \gtrsim 2$) resulted in a remarkable enhancement of the activity. Propylene polymerization was then carried out with the soluble $Ti(OBu)_4/AlEt_2Cl$ catalyst combined with a variety of metal chlorides (MCl_x). A clear relation was found between the activity and the electronegativity (X_i) of metal ion in MCl_x. The activity drastically changed around the X_i of Ti^{3+} ($X_i \cong 10.5$) : MCl_x containing metal ions with X_i below 9 ($MgCl_2$, $CoCl_2$, $MnCl_2$ etc.) accelerated the activity, whereas those with X_i above 11 ($AlCl_3$, $ZrCl_4$, $NaCl_5$ etc.) retarded it. Such a correlation was also observed in the polymerization of ethylene with the use of soluble chromium-based catalytic systems. From these results combined with the XPS spectra of Cl(2p) in MCl_x and of Ti(2p) in $TiCl_4/MCl_x$, a plausible mechanism was proposed for the enhancement of the activity by using $MgCl_2$.

Introduction

The activity of Ziegler-Natta catalysts for olefin polymerization has been substantially increased by supporting the titanium compound on a solid carrier.[1-23] A variety of metal halides have been reported as carriers, the most common being magnesium chloride ($MgCl_2$). The first example of such a supported $TiCl_4$ catalyst dates back to 1955, although the patent[24] only gives example of ethylene polymerization. Early Shell patent applications in this area include a filing in 1960 describing the first $MgCl_2$-supported $TiCl_4$ catalyst for propylene polymerization.[25] The patent describes the preparation of catalysts by supporting $TiCl_4$ on carriers such as $MgCl_2$ and $CoCl_2$ which could be ground prior to use. The preferred halides were those having the same lattice structure as δ-$TiCl_3$ catalysts. Furthermore, the use of electron

R. B. Seymour and T. Cheng (eds.), History of Polyolefins, 243–256.

donors, notably the esters of carboxylic acids was demonstrated to increase stereoselectivity. After these initial patents there followed many patents in the 1960's and 1970's describing $MgCl_2$-supported $TiCl_4$ catalysts for olefin polymerization.

The remarkable improvement of the activity by using $MgCl_2$ has recently been revealed to be attributed to a drastic increase in the propagation rate constant as well as a considerable increase in the number of active species.[16-20] However, the precise reason why $MgCl_2$ peculiarly accelerates the propagation rate is still quite open to discussion.

From such a viewpoint, the influence of $MgCl_2$ on the activity was examined in the polymerization of ethylene and propylene by using both titanium- and chromium-based catalysts.

Experimental

Materials

Research grade ethylene and propylene (from Takachiho Chemical Co.) were purified by passing through the molecular seive 3A column dried at 350°C for 50 h. Nitrogen of ultra high purity (from Nippon Sanso Co. 99.9989 % purity) was further purified by passing through the molecular seive 3A column. SiO_2 (from Fuji Davison Co., 300 m^2g^{-1}) was evacuated at 300°C for 4 h immediately before use. Commercial anhydrous metal chlorides (MCl_x), such as $MgCl_2$, $MnCl_2$, $CoCl_2$ etc., were further dried under vacuum at 200 - 400°C for 5 h. Research grade heptane (from Wako Pure Chemical Co.) was purified according to the usual procedure. The other chemicals (research grade) were commercially obtained and used without further purification.

Preparation of $MgCl_2/TiCl_4/SiO_2$ Catalyst

To a 200 ml glass reactor equipped with a mangetic stirrer were added 90 ml of heptane, 10 g of SiO_2 and 40 ml of $TiCl_4$ under nitrogen. The mixtures were kept standing with vigorous stirring for 6 h at 98.4°C. To the reactor containing the precipitates was added 40 ml of heptane, and the mixtures were again kept standing with vigorous stirring for 6 h at r.t. to remove the unreacted $TiCl_4$. The precipitates were separated by filtration under nitrogen, adequately washed with heptane and dried under vacuum to give the SiO_2-supported $TiCl_4$ catalyst. Then, to 40 ml of heptane containing 1.0

g of the SiO_2-supported $TiCl_4$ catalyst and 20 mmol of $AlEt_2Cl$ was added the measured amount of $MgCl_2(THF)_2$, which had been prepared by dissolving $MgCl_2$ in THF followed by evacuation at r.t. for 5 h. The mixtures were kept standing with vigorous stirring at r.t. for 20 h. The precipitates were separated by filtration, washed with plenty of heptane and dried under vacuum at r.t. The catalyst showed green color, which implies the formation of a complex between $MgCl_2$ and the titanium species.

Preparation of $Ti(OBu)_4/MCl_x$ Catalyst

10 mmol of each MCl_x was dissolved in 90 mmol of 2-ethyl-1-hexyl-alcohol (2-EHA) by refluxing the solvent, and the solution was diluted to 0.047 mmol/l in heptane and reserved as a stock solution. Immediately before polymerization, 0.085 mmol of $Ti(OBu)_4$ and a prescribed amount of the stock solution of MCl_x were introduced to the polymerization vessel under nitrogen, and the mixtures were kept standing with vigorous stirring for 15 min at the polymerization temperature.

Preparation of Soluble Chromium-based Catalysts

$Cr(CH_3COO)_3 \cdot H_2O$ was treated in the mixture of acetic anhydride and acetic acid at 120°C for 4 h, and the excess solvent was removed under reduced pressure to give a bright green fine powdery product, $Cr(CH_3COO)_3 \cdot (CH_3CO)_2O$. 0.84 mmol of the product was dissolved in 167 mmol of 2-EHA.

$Cr(OBu)_4$ was prepared according to the method reported by Bradly et al.[26] from BuOH and $Cr(NEt_2)_3$, followed by oxygenation. This compound was freely soluble in toluene.

Polymerization and Analytical Procedures

The polymerization was carried out in the same apparatus and procedures reported earlier.[9] The rate of polymerization was determined from the rate of monomer consumption measured by a hot-wire flowmeter with a recorder. Some experiments were conducted in a stainless steel reactor.

The catalysts were analyzed by atomic absorption spectrophotometery (Shimadzu AA-6105) and elemental analysis.

XPS spectra were recorded with a Shimadzu electron spectrometer ESCA 750 with MgK_α excitation radiation. A glove box was attached to the ESCA spectrometer and filled with nitrogen. Sample preparation for ESCA measurements was performed in the box. The pressure in the spectrometer was about 2-3×10^{-7} Torr. The contamination C (1S) line was taken as an internal standard at 285.5 eV. The $TiCl_4/MCl_x$ samples were prepared by grinding the mixtures of $TiCl_4$ and MCl_x in a ball mill under nitrogen at r.t. The ESR spectra were recorded with a Varian E-12 ESR spectrometer at 20°C at a microwave frequency of 9.528 GHz.

Results and Discussion

Polymerization of Propylene with Ti-based Catalysts

The majority of highly active catalysts reported so far have been prepared by supporting the titanium species on the surface of $MgCl_2$. On the other hand, Union Carbide Corp.[27] has recently developed a new type of catalyst by supporting both titanium compound and $MgCl_2$ on a solid carrier. The catalyst contains less amount of chloride, which eliminates the procedure to remove a catalyst residue from the products. Besides such an industrial importance, the catalyst seems to give us a useful information as to why $MgCl_2$ markedly accelerates the propagation rate constant. In this connection, a well-defined such type of catalyst ($MgCl_2/TiCl_4/SiO_2$) was prepared according to the procedures described in the experimental section.

Table 1

Results of propylene polymerization[a].

Run No.	Catalyst	Mg/Ti	Activity (kg-pp/g-Ti·h)
1	$TiCl_4/SiO_2$	0	0.73
2	$TiCl_4/MgCl_2/SiO_2$	1.2	4.09
3	"	1.8	14.8
4	"	2.2	11.9
5	"	3.1	12.5
6	"	4.2	13.9
7	"	4.6	10.6

a) Polymerization was conducted in a stainless steel reactor at 65°C for 15 min by using 0.1 mmol of $AlEt_3$, 10 cm^3 of heptane and 11 g of propene. Ti supported: 0.51 matom/g-cat.

Typical results of propylene polymerization obtained by using the catalyst combined with $AlEt_3$ are shown in Table 1. With an increase in the molar ratio of Mg/Ti the activity drastically increased to attain a constant value at the molar ratio of about 2.

To obtain a better insight into the feature of active species, the $MgCl_2/TiCl_4/SiO_2$ catalyst was washed with THF under nitrogen at r.t. Table 2 shows the analytical results of the catalyst before and after washing with THF. Most of the $MgCl_2$ was removed by this procedure, which indicates that a week complex is formed between $MgCl_2$ and the titanium species.

Table 2

Analytical results of the $MgCl_2/TiCl_4/SiO_2$ catalyst[a).]

Ti Supported[b)] (matom/g-cat.)	Mg Supported[b)] (matom/g-cat.)	Cl Supported[c)] (matom/g-cat.)
0.51[d)]	1.57	—
0.46[e)]	0.003	0.95

a) The catalyst used in Run No. 5 in Table 1 was employed.
b) Measured by atomic absorption spectrophotometry.
c) Measured by elemental analysis.
d) before washing.
e) after washing.

In Figure 1 is shown a plausible scheme for the formation of the active species.[28)]

Polymerization of propylene was then carried out with the soluble $Ti(OBu)_4/MgCl_2/AlEt_2Cl$ catalytic system by changing the molar ratio of Mg/Ti. This system gave a stationary rate after a short induction period. The stationary rate markedly increased with an increase in the molar ratio of Mg/Ti to attain a constant value at about 6.[29)] This result together with that abtained with the $MgCl_2/TiCl_4/SiO_2$ catalyst clearly indicate that a remarkable enhancement in the polymerization activity is attained by using only a small amount of $MgCl_2$.

Propylene polymerization was subsequently carried out under similar conditions by using various types of MCl_x ($[MCl_x]/[Ti] = 6$) in place of $MgCl_2$.[29)]

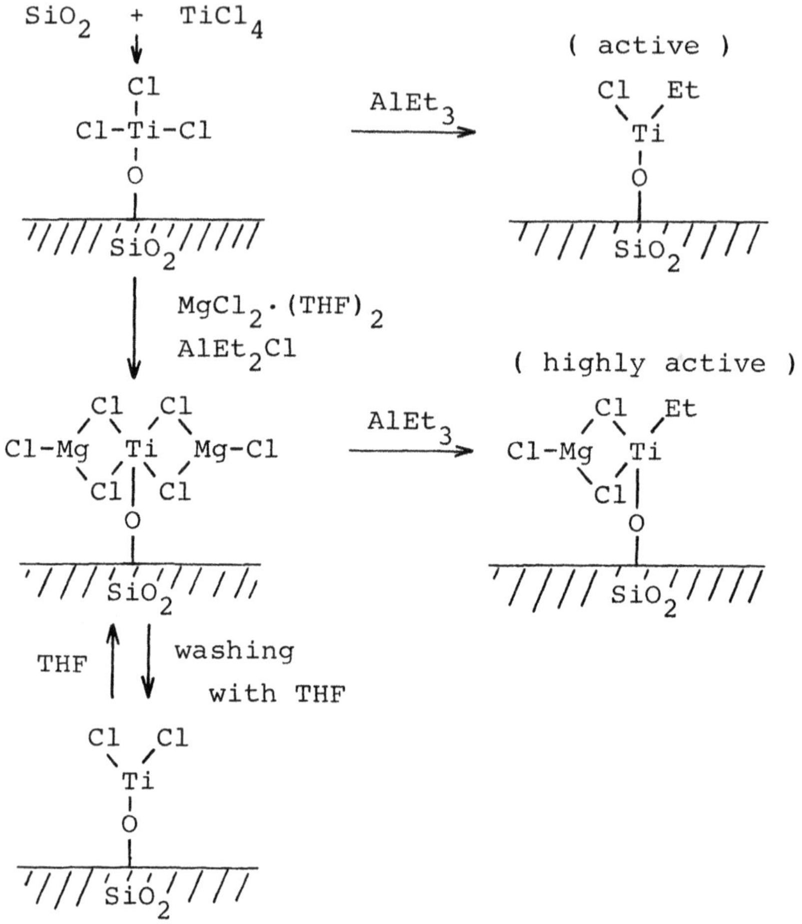

Figure 1

Plausible mechanism for the formation of active species with $TiCl_4/SiO_2/MgCl_2$

Every polymerization run showed an individual stationary rate.

In Figure 2 is shown the correlation between the stationary rate and the electronegativity (X_i) of the metal ion in MCl_x. X_i was calculated from the empirical equation derived by Tanaka et al.[30)]

The results shown in Figure 2 clearly indicate that the metal chlorides of $X_i \lesssim 9$ act as accelerators whereas those of $X_i \gtrsim 11$ do as inhibitors.

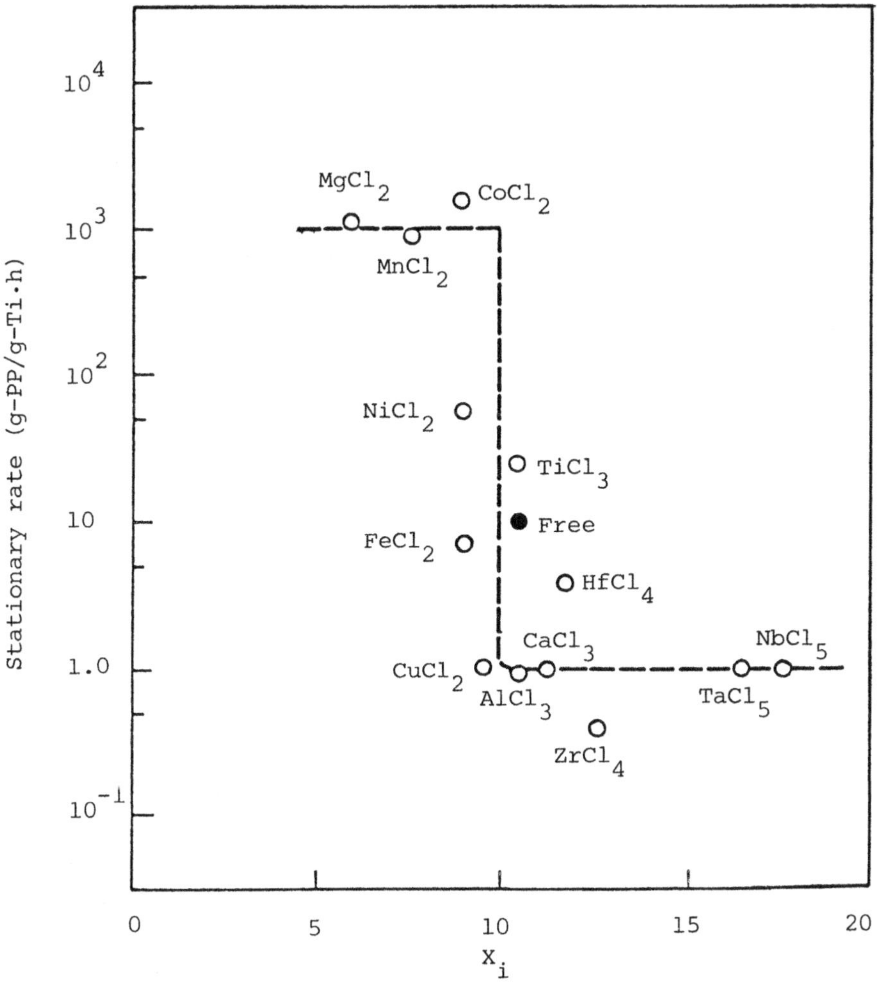

Figure 2

Relationship between stationary rate and electronegativity X_i of metal ion in MCl_x.
Polymerization conditions: 41°C, propylene = 1 atm, $Ti(OBu)_4$ = 0.085 mmol, DEAC/Ti = 100, MCl_x/Ti = 6, heptane = 200 ml.

Escalona[31] has also found that the metal halides [NaCl: $X(Na^+) = 2.7$, $BaCl_2$:$X(Ba^{2+}) = 4.5$, $SrCl_2$:$X(Sr^{2+}) = 5.0$, $CaCl_2$:$X(Ca^{2+}) = 5.0$, $MgCl_2$:$X(Mg^{2+}) = 6.0$, $MnCl_2$:$X(Mn^{2+}) = 7.5$] modification of the SiO_2 followed by support of $TiCl_4$ has a pronounced effect on the activity of ethylene polymerization. These results seem to be in good agreement with our data.

Then, the apparent activation energy of the stationary polymerization rate was measured with the catalytic systems of $Ti(OBu)_4/MgCl_2$, $Ti(OBu)_4/MnCl_2$, $Ti(OBu)_4/CoCl_2$ and MCl_x-free $Ti(OBu)_4$. The apparent activation energy decreased in the following order.

$Ti(OBu)_4$ (14.0±0.2 kcal/mol) > $Ti(OBu)_4/MnCl_2$ (13.0±0.1)

> $Ti(OBu)_4/MgCl_2$ (12.2±0.1) > $Ti(OBu)_4/CoCl_2$ (11.5±0.3)

The increase in the activity by employing these metal chlorides seems mainly attributed to a decrease in the activation energy of polymerization [exp[(14000-11500)/RT] ≑ 65].

This result strongly suggests that the promotion by these metal chlorides is mainly caused by electron donation to the active titanium(Ti^{3+}). From such a point of view, XPS spectra were recorded on various metal chlorides (MCl_x) as well as several $TiCl_4/MCl_x$ systems.[32]

Table 3

XPS analysis of MCl_x.

Metal chloride	Binding energy of Cl_{2p} (ev)	X_i
$MgCl_2$	200.1	6.0
$MnCl_2$	200.1	7.5
$CoCl_2$	200.1	9.0
$TiCl_3$	200.6	10.5
$AlCl_3$	200.8	10.5
$HfCl_4$	200.4	11.7
$ZrCl_4$	200.8	12.6
$TaCl_5$	200.5	16.5

In Table 3 are shown the binding energies (E_B) of Cl (2p) of MCl_x together with X_i of the metal ions in MCl_x. The electron densities of Cl (2p) in the metal chlorides of Xi$\lesssim$9 seem to be considerably rich as compared with those of Xi$\gtrsim$11. Table 4 shows the E_B values of Ti (2p3/2) and Ti (2p1/2), which implies that the electron densities of Ti(2p) of $TiCl_4/MgCl_2$ and $TiCl_4/MnCl_2$ are also rich to some extent in comparison to those of $TiCl_4/AlCl_3$ and $TiCl_4/ZrCl_4$.

Table 4

XPS analysis of $TiCl_4/MCl_x$ catalyst.

Catalyst	Binding energy of Ti (ev) $Ti_{2p3/2}$	$Ti_{2p1/2}$	X_i
$TiCl_4/MgCl_2$	459.8	465.4	6.0
$TiCl_4/MnCl_2$	459.5	465.3	7.5
$TiCl_4/AlCl_3$	460.5	466.0	10.5
$TiCl_4/ZrCl_4$	460.5	466.2	12.6

In the ammonia synthesis reaction, it is well known that transition metals are activated by alloying with electropositive metals. For example, Raney Ru prepared from Ru-Al alloy is more active than pure Ru in terms of specific activity per surface Ru, suggesting promotion by residual Al remaining after leaching, while it is further activated by addition of alkali metals to give a highly active catalyst. Ozaki et al. [33] has developed a theory of promotion in which the activation of Ru is attributted to an increase in electron concent ration which strengthens the Ru-N bond simultaneously weakening the N≡N bond.

The present results may be interpreted in terms of a similar mechanism as described in the following. The metal chlorides are involved in the active species and influence the electronic structure of the active titanium (Ti^{3+})[9] by inductive effects. The electron-donating capability of $MgCl_2$, $MnCl_2$ or $CoCl_2$ to active Ti^{3+} stabilizes the coordination of a propylene monomer with Ti^{3+} if the back-donation of electron from Ti^{3+} to propylene is the main factor in determining a titanium-propylene interaction, resulting in an acceleration of a subsequent insertion of the coordinated propylene into an active titanium-polymer chain bond. In contrast, the electron-withdrawing capability of metal chlorides may cause an unfavourable effect in the chain propagation. A plausible model is shown in Figure 3.

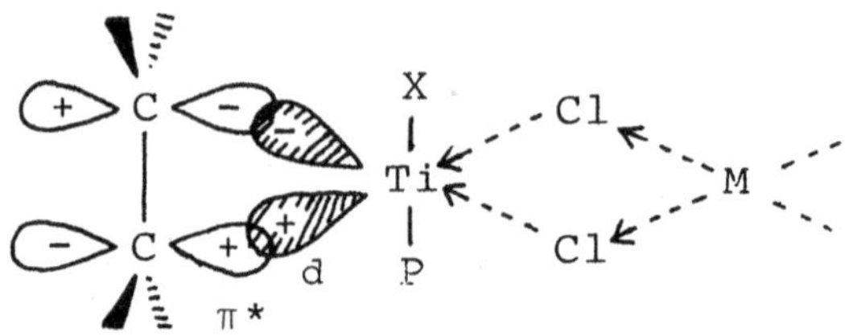

Figure 3

Plausible model of MCl_x-promotion of Ti^{3+}.

Polymerization of Ethylene with Cr-based Catalysts

Polymerization of ethylene was first conducted at 30°C in toluene with the use of $Cr(CH_3COO)_3 \cdot (CH_3CO)_2O$ and $Cr(OBu)_4$ catalysts by employing $AlEt_3$, $Al_2Et_3Cl_3$ and $AlEt_2Cl$ as cocatalyst. The activity was markedly dependent upon the cocatalyst used and $AlEt_2Cl$ showed the highest activity among these alkylaluminums. From the ESR analysis of the catalytic systems combined with a detailed kinetic study of ethylene polymerization, the active species in these systems was found to be Cr^{2+}.[34,35]

The addition of $MgCl_2$ to the $Cr(CH_3COO)_3 \cdot (CH_3CO)_2O/AlEt_2Cl$ system resulted in a remarkable increase in the activity (Figure 4). While, no apparent enhancement was observed when $AlEt_3$ was used in place of $AlEt_2Cl$, indicating that the ligands exchange reaction takes place between $Cr(CH_3COO)_3 \cdot (CH_3CO)_2O$ and $AlEt_2Cl$, followed by forming an active complex between the resultant Cr-compound and $MgCl_2$ through Cl bridges. Similar activation process is expected to proceed also in the case of the $Ti(OBu)_4/AlEt_2Cl/MCl_x$ systems.

The polymerization of ethylene was subsequently conducted with the $Cr(CH_3COO)_3 \cdot (CH_3CO)_2O/AlEt_2Cl$ and $Cr(OBu)_4/AlEt_2Cl$ systems by using various metal chlorides in place of $MgCl_2$. In Figure 5 is shown the correlation between the activity and the electronegativity of metal ions in MCl_x with the former system. Similar results were obtained with the use of the latter system.[36] These results can also be interpreted by the same mechanism as proposed above in the propylene polymerization with the use of $Ti(OBu)_4/AlEt_2Cl/MgCl_x$ systems.

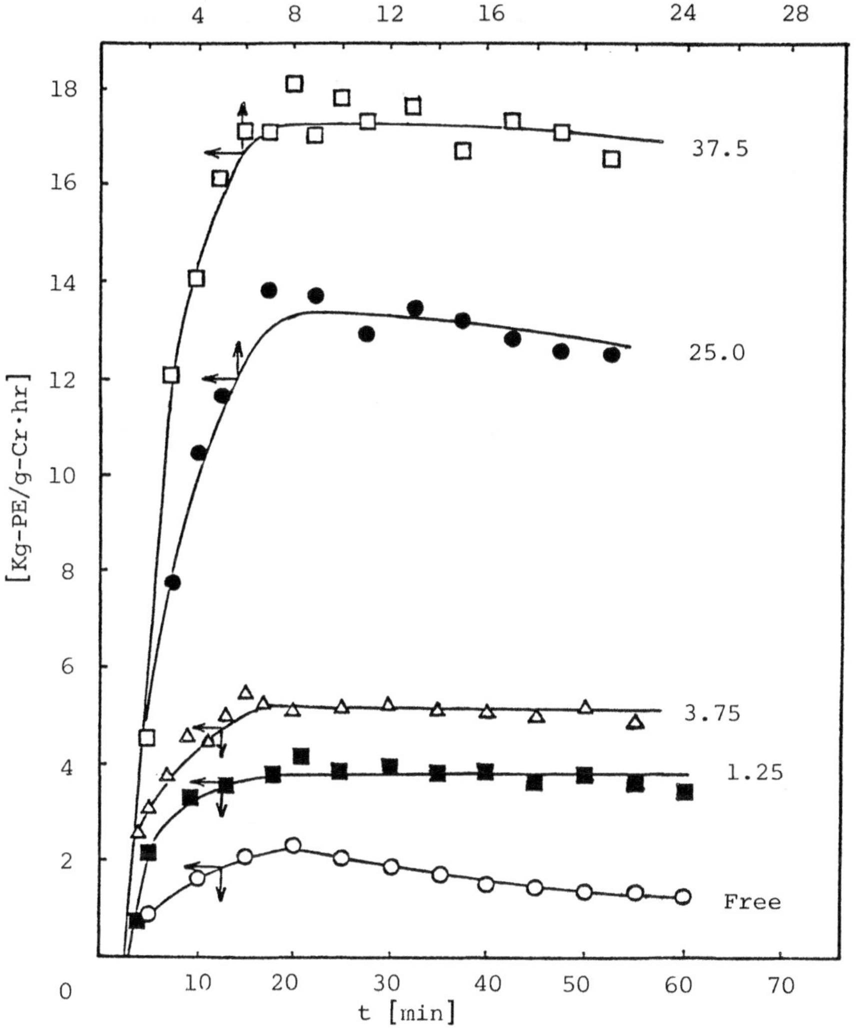

Figure 4

Effect of $MgCl_2$ on the rate of ethylene polymerization. Polymerization conditions: 30°C, ethylene = 1 atm, Cr compd. = 5.3 mg, DEAC/Cr = 10^3, toluene = 100 ml.

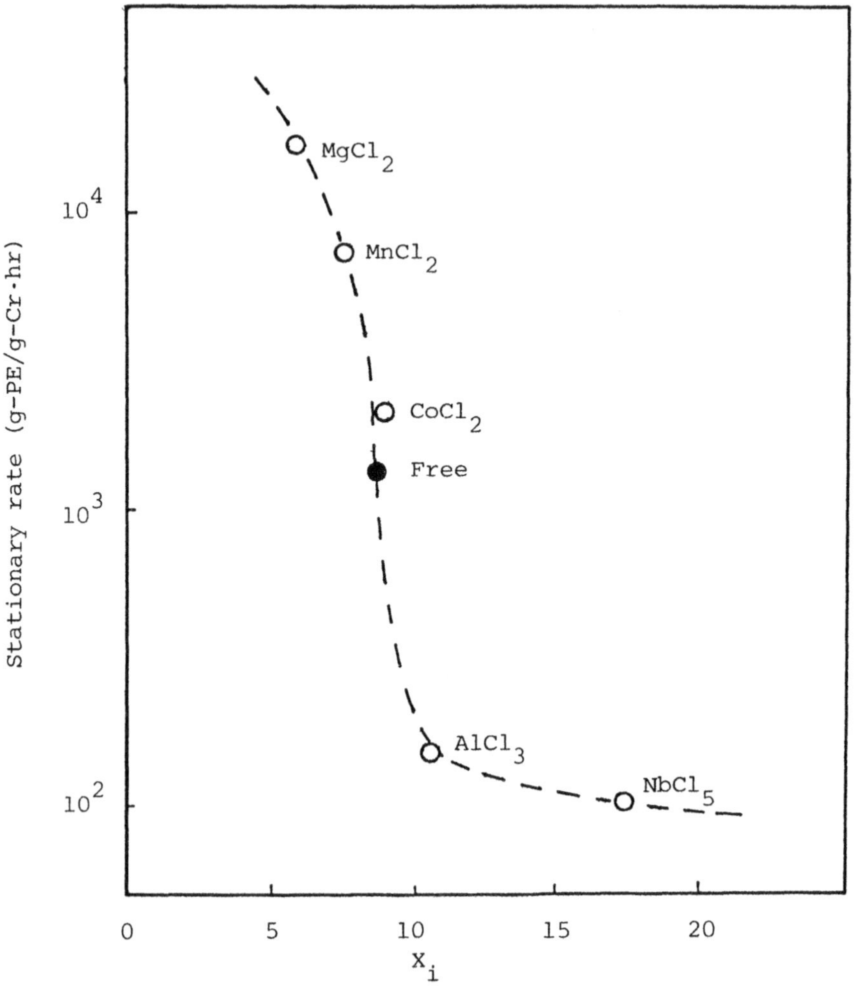

Figure 5

Relationship between stationary rate and electronegativity X_i of metal ion in MCl_x.
Polymerization conditions: 30°C, ethylene = 1 atm, Cr compd. = 5.3mg, DEAC/Cr = 10^3, MCl_x/Cr = 40, toluene = 100 ml.

Acknowledgements

I would like to express my thanks to Professor Emeritus Tominaga Keii for his continuing guidance over two decades and also his kind encouragement. I am pleased to thank all of my collaborators cited in the references section. I also acknowledge the contribution of Dr. N.Tomotsu of Petrochemical Research Division, Idemitsu Kosan Co. Ltd. for obtaining the XPS spectra.

Kazuo Soga

Research Laboratory of Resources Utilization, Tokyo Institute of Technology, 4259 Nagatsuta-cho, Midori-ku Yokohama 227 Japan

References

1) W.A. Hewett, J. Polym. Sci., B, 5, 855 (1965)
2) K. Soga, S. Katano, Y. Akimoto, T. Kagiya, Polym. J., 5, 128 (1973)
3) R.N. Haward, A.N. Koper, K.L. Fletcher, Polymer, 14, 365 (1973)
4) J.C.W. Chien, J.T.T. Hsieh, J. Polym. Sci., Polym. Chem. Ed., 14, 1915 (1976)
5) A. Munoz-Escalona, J. Villalba, Polymer, 18, 179 (1977)
6) K. Gardner, I.W. Parsons, R.N. Haward, J. Polym. Sci., Polym. Chem. Ed., 16, 1983 (1978)
7) E. Suzuki, M. Tamura, Y. Doi, T. Keii, Makromol. Chem., 180, 2235 (1979)
8) E.W. Duck, D. Grant, E. Kronfli, Eur. Polym. J., 15, 625 (1979)
9) K. Soga, T. Sano, R. Ohnishi, Polym. Bull., 4, 157 (1981)
10) K. Soga, K. Izumi, M. Terano, S. Ikeda, Makromol. Chem., 181, 657 (1980)
11) S.S. Ivanchev, A.A. Baulin, A.G. Rodionov, J. Polym. Sci., Polym. Chem. Ed., 18, 2045 (1980)
12) N. Kashiwa, Polym. J., 12, 603 (1980)
13) H.L. Hsieh, Polym. J., 12, 596 (1980)
14) K. Soga, M. Terano, Polym. Bull., 4, 39 (1981)
15) P. Galli, L. Luciani, G. Cecchin, Angew. Makromol. Chem., 94, 63 (1981)
16) Y. Giannini, Makromol. Chem. Suppl., 5, 216 (1981)

17) T. Keii, E. Suzuki, M. Tamura, M. Murata, Y. Doi, Makromol. Chem., 183, 2285 (1982)
18) Y. Doi, M. Murata, K. Yano, T. Keii, Ind. Eng. Chem. Prod. Res. Dev., 21, 580 (1982)
19) N. Kashiwa, J. Yoshitake, Makromol. Chem., Rapid Commun., 3, 211 (1982); Mokromol. Chem., Rapid Commun., 4, 41 (1983)
20) G.D. Bukatov, S.H. Shepelev, V.A. Zakharov, S.A. Sergeev, Yu.I. Yermakov, Makromol. Chem., 183, 2657 (1982)
21) J.C. W. Chien, J.C. Wu, I.J. Kuo, J. Polym. Sci., Polym. Chem. Ed., 20, 2091 (1982)
22) A.D. Caunt, J.A. Licchelli, I.W. Parsons, R.N. Haward, M.R.Y. AI-Hillo, Polymer, 24, 121 (1983)
23) P.C. Barbe, L. Noristi, G. Baruzzi, E. Marchetti, Makromol. Chem., Rapid Commun., 4, 249 (1983)
24) Petrochemicals Ltd., Brit. Pat. 841, 822 (1955)
25) Shell, Brit. Pat. 904, 510 (1960)
26) E.C. Alyea, J.S. Basi, D.C. Bradley , M.H. Chisholm, J. Chem. Soc. (A), 1971, 772
27) Union Carbide Corp., Japan Kokai, 54-142192 (1979)
28) K. Soga, R. Ohnishi, Y. Doi, Polym. Bull., 9, 299 (1983)
29) Y. Doi, K. Soga, M. Murata, E. Suzuki, Y. Ono, T. Keii, Polym. Commun., 24, 244 (1983)
30) K. Tanaka, A. Ozaki, K. Tamaru., Shokubai, 6, 262 (1984)
31) A. Munoz-Escalona, 'Transition Metal Catalyzed Polymerizations', R.P. Quirk ed., Harwood Academic Publishers, New York, Part A, p323 (1983)
32) K. Soga, Y. Doi, T. Keii, Proc. 8th Intern. Congr. Catalysis, West Berlin, Vol.5, p349 (1984)
33) A. Ozaki, K. Aika, 'Catalysis-Science and Technology', J.R. Anderson, M. Boudart eds., Springer-Verlag, Berlin, 1981, Vol.1, p87.
34) K. Soga, S.I. Chen, T. Shiono, Y. Doi, Polymer, in press
35) S.N. Gan, S.I. Chen, R. Ohnishi, K. Soga, Makromol. Chem., Rapid Commun., 5, 535 (1984)
36) K. Soga, S.I. Chen, T. Shiono, Y. Doi, Polymer, in press

W. Kaminsky

SYNTHESIS OF POLYOLEFINS WITH HOMOGENEOUS ZIEGLER-NATTA-CATALYSTS OF HIGH ACTIVITY

1. INTRODUCTION

For a long time homogeneous Ziegler-catalysts basing on bis(cyclopentadienyl)titanium or zirconium dialkyls and trialkylaluminiums were assumed to be inactive for olefin polymerization. In the case of biscyclopentadienyl-ethyl-titanium chloride and triethylaluminium very fast aging reactions take place which end up before long with titaniumIII-compounds. These titaniumIII-compounds are as inactive as mixtures of biscyclopentadienyltitaniumdimethyl and trimethylaluminium.

An accident in our laboratory in 1976 brought about equimolar amounts of water into the system compared to the trimethylaluminium, and, surprisingly, an unusual high polymerization activity of ethylene was observed. Activities of more than 10,000gPE/g Ti in two hours were obtained. If the aluminium to water ratio ($Al(CH_3)_3$:H_2O) drops below 1/3 all methyl groups are hydrolyzed and the activity disappears. The study of the influence of water in the following time led to the isolation of oligomeric alumoxanes varying in the extent of condensation and structure element

$$\left[\begin{array}{l} CH_3 \\ | \\ Al{-}O \end{array}\right]_n \qquad n = 6\text{-}20$$

The alumoxanes formed by reaction of water and trimethylaluminium under elemination of methane are a mixture of linear and cyclic structures.

2. POLYETHYLENE

Using separately prepared alumoxanes together

R. B. Seymour and T. Cheng (eds.), History of Polyolefins, 257–270.

with zirconium compounds an extremely high polymerization activity is found (Table I). Especially with temperatures over 50 °C the zirconium catalyst gives higher activities than the titanium or hafnium system. Of the cocatalysts methylalumoxane is more useful than ethylalumoxane or isobutylalumoxane[1-3].

Table I

Polymerization activity of the bis(cyclopentadienyl)zirconiumdimethyl/methylalumoxane catalyst applied to ethylene.

activity (90 °C, 8 bar)	=	$24,8 \cdot 10^6$ g PE/g Zr·h
(Zr)	=	$3,3 \cdot 10^{-8}$ mol/l
$(Al(CH_3)-O)$	=	$2,2 \cdot 10^{-3}$ mol/l
molecular weight of the obtained polyethylene	=	150,000
degree of polymerization	=	5,360
macromolecules per Zr atom and h	=	15,000
rate of growth of one macromolecule	=	0,27 sec
turnover time	=	$5 \cdot 10^{-5}$ sec

Assuming that every zirconium atom forms an active complex one zirconium complex produces about 15,000 polymer chains in one hour. The turnover time of $5 \cdot 10^{-5}$ sec is of the same value as that of many enzymes.

Many differen+ zirconocenes are active for olefin polymerization in combination with alumoxane. Fig. 1 shows the polymerization rate versus time diagram at constant temperature and concentrations.

The maximum activity is reached after 5 to 10 minutes. Bis(cyclopentadienylzirconium)dichloride is most active while bis(pentamethylcyclopentadienyl)zirconiumdichloride gives the lowest rate. The molecular weight M_η of the obtained polyethylene lies between 115-120,000.

The properties of the polyethylenes obtained are remarkable (Table II).

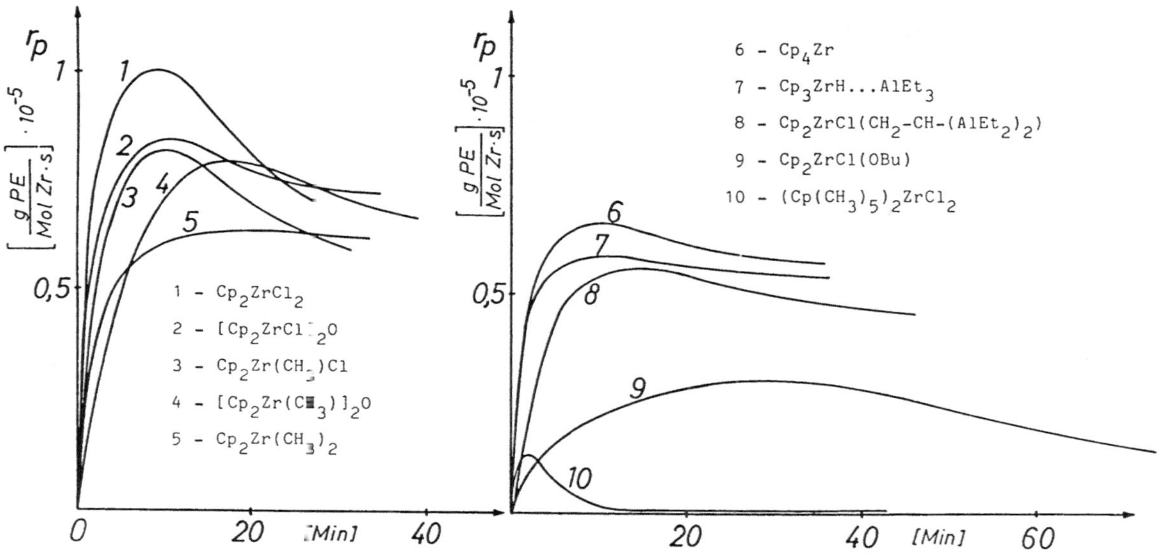

Fig. 1: Polymerization of ethylene with different zirconocenes catalysts, activity versus time, polymerization temperature 80 °C, (Zr) = 10^{-7} mol/l, methylaluminoxane concentration $1{,}46 \cdot 10^{-2}$ mol Al/l, ethylene pressure 1,5 bar, solvent = 330 ml toluol

Table II

Density and melting point of polyethylene samples with different molecular weights.

M_w	M_w/M_n	density of original material g/cm^3	density of pressed material g/cm^3	melting point ^oC
200 000	2,0	0,98	0,947	140,5
147 000	1,9	0,973	0,953	140,0
100 000	1,8	0,97		139,0

The density of the raw polyethylene powder is 0,98 g/cm^3 and decreases to 0,95 g/cm^3 for pressed plates. Density and melting point decrease with decreasing molecular weight. The molecular weight distribution M_w/M_n is narrow.

3. REGULATION OF MOLECULAR WEIGHT AND DENSITY

Regulation of the average molecular weight in this homogeneous Ziegler system can be achieved by either varying the concentration of Cp_2ZrCl_2, changing the reaction temperature or adding small amounts of comonomers. With increasing zirconium concentration the molecular weight decreases nearly linearily. This leads to the conclusion that for the chain transfer a bimetallic mechanism is necessary.

At a reaction temperature of 10 ^oC, polyethylene is formed with a molecular weight of 1,500,000, whereas at 50 ^oC it has descending to 180,000 only to go down to 90,000 at 90 ^oC. Raising the temperature to more than 100 ^oC yields α-olefins with an even number of carbon atoms.

The molecular weight can also be influenced by addition of hydrogen (Fig. 2).

In contrast to most heterogeneous catalysts, here only traces of hydrogen are needed to lower the molecular weight over a wide range. To reduce the molecular weight to about one half of that

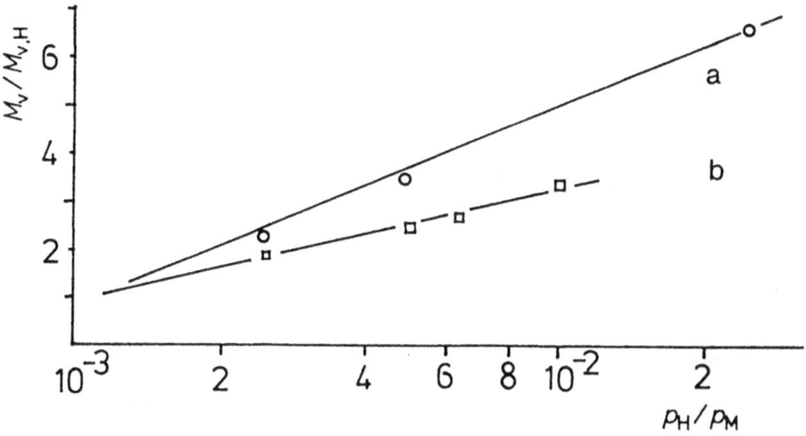

Fig. 2: Relationship between the ratio of viscosity molecular weight M_V obtained without hydrogen $M_V/M_{V,H}$ and the ratio ρ_H/ρ_M of the concentration of hydrogen and ethylene of the mixture in the feed. (Al) = $1{,}72\cdot10^{-2}$ mol/l. Curve a: (Cp_2ZrCl_2) = $2\cdot10^{-5}$ mol/l, 60 °C, curve b: (Cp_2ZrCl_2) = $7{,}5\cdot10^{-6}$ mol/l, 70 °C.

value obtained without hydrogen, it is necessary to feed the reactor only with 7,5 vol.-% H_2.

The density of the polymer can be lowered by using hexene-1 or butene-1 as an additional monomer in copolymerizations. Incorporation of 1 mol % hexene-1 yields a polymer which, when in the molten state, has a density lower than 0,95 g/cm^3. With 5 mol % incorporated the density is 0,92 g/cm^3 and decreases to 0,89 g cm^3 for 9 mol %, so LLDPE can be obtained.

4. POLYETHYLENE WITH FILLING MATERIALS

As a result of the increasing costs of raw materials nowadays more efforts are made to incorporate biomass into the industrial production. Natural substances like starch, cellulose and lignin can, due to their free water, form alumoxane analogous structures on addition of trimethylaluminium. After addition of the metal compound starch grains or cellulose fibers treated in this manner have a considerable polymerization activity when suspended in toluene (Fig. 3).

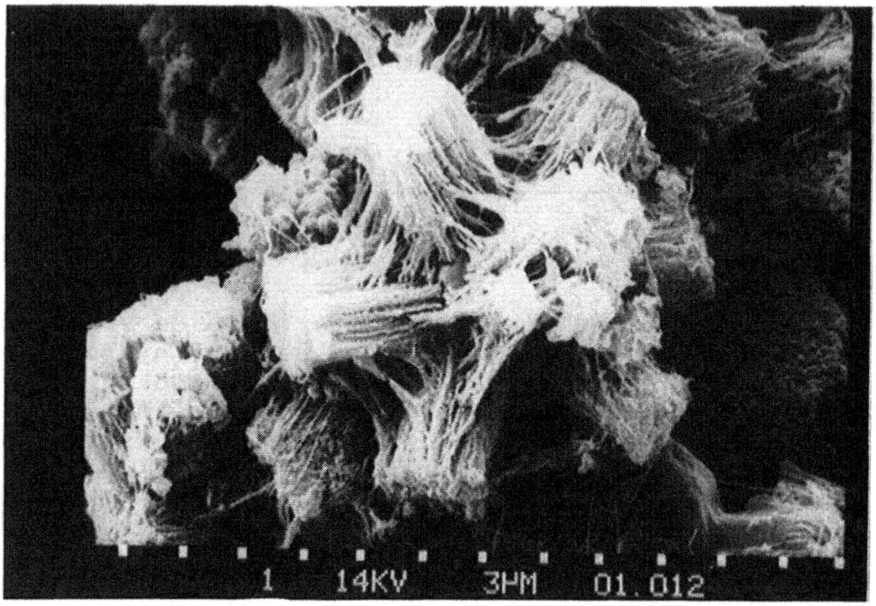

Fig. 3: Electron microscope photograph of starch grains which are covered with polyethylene.

Fig. 3 shows a starch grain on which polyethylene crystals of 1 to 5 μ in length have grown[7].

By this procedure physical and chemical properties of natural and synthetic polymers can be combined.

In the same manner it is possible to cover cellulose, lignin or inorganic materials like $CaCO_3$, $CaSO_4 \cdot 0{,}5\ H_2O$ or Al_2O_3 with polyethylene or other polyolefins.

5. POLYPROPYLENE

These catalysts are also able to polymerize propylene and to copolymerize ethylene and propylene

Table III

Polymerization of propylene with the catalyst $Cp_2Zr(CH_3)_2$/Methylalumoxane

Zr mol	Al mol	Toluene ml	Propylene ml	T °C	Time h	Yield g	Activity $\frac{kg\ PP}{g\ Zr \cdot h}$	$M_\eta \cdot 10^{-3}$
$5{,}1 \cdot 10^{-6}$	$7{,}8 \cdot 10^{-3}$	50	415	5	20	219,5	23,4	75
$1{,}7 \cdot 10^{-6}$	$5{,}0 \cdot 10^{-3}$	-	330	5	24	48,5	13,3	106
$3{,}3 \cdot 10^{-6}$	$5{,}0 \cdot 10^{-3}$	-	330	5	24	101,0	13,9	82
$3{,}5 \cdot 10^{-6}$	$5{,}4 \cdot 10^{-3}$	5	340	5	24	120,0	15,7	106
$3{,}5 \cdot 10^{-6}$	$5{,}4 \cdot 10^{-3}$	5	340	-10	12	32,0	8,4	191
$3{,}6 \cdot 10^{-6}$	$5{,}9 \cdot 10^{-3}$	10	330	-30	11	6,3	1,7	
$3{,}5 \cdot 10^{-5}$	$5{,}9 \cdot 10^{-3}$	50	250	-30	50	202,2	1,3	
$2{,}0 \cdot 10^{-5}$	$3{,}4 \cdot 10^{-3}$	50	100	-70	58	8	0,07	590

After passing a maximum, a constant rate of polymerization is observed. Only atactic polypropylene is formed with Cp_2ZrIV-compounds, $Cp(CH_3)_5$-$CpZrCl_2$ or $(Cp(CH_3)_5)_2ZrCl_2$. For low temperatures only Ewen[4] reported that small amounts if isotactic polypropylene are obtained with Cp_2ZrPh_2. Even the atactic polypropylene shows a narrow molecular weight distribution of 1,6 to 2,5.
The zirconium catalyst provides the way to ethylene/propylene copolymers (EPM). Changing the solvent, for example toluene or liquid propylene, and the ethylene pressure is a means by which the desired ratio of propylene or other olefins to ethylene in the polymer can be regulated. As a result, the properties of the copolymerization products can be influenced considerable. Copolymerization products with 10 to 90 % of propylene are yielded.

EPDM terpolymers with ethylidene norbornene as diene monomer could also be prepared with $Cp_2Zr(CH_3)_2$/methylalumoxane (Table 4).

Table IV

EPDM - polymerization conditions, yields and activities

C_2H_4 bar	C_3H_6 bar	Diene g/ml x 10^2	Cp_2ZrMe_2 mol/l	T °C	t h	Activity $\frac{\text{kg EPDM}}{\text{molZr}\cdot\text{h}\cdot\text{bar}}$
1.5	7.5	1.4	$1\cdot10^{-6}$	20	24	225.3
1.5	7.5	1.0	$5\cdot10^{-7}$	10	20	213.9
3.3	5.7	1.1	$1\cdot10^{-6}$	20	4	1527.7
2.5	6.5	2.2	$1\cdot10^{-5}$	20	6	140.0
3.3	5.7	1.5	$1\cdot10^{-6}$	5	48	38.9
4.0	5.0	1.5	$1\cdot10^{-6}$	20	3	158.7

Overall activities at 20 °C lie between 100 and 1,500 kg EPDM/mol Zr h bar. Even at low transition metal concentrations of less than 10^{-6} mol/l the yields are responsible (Fig. 4).

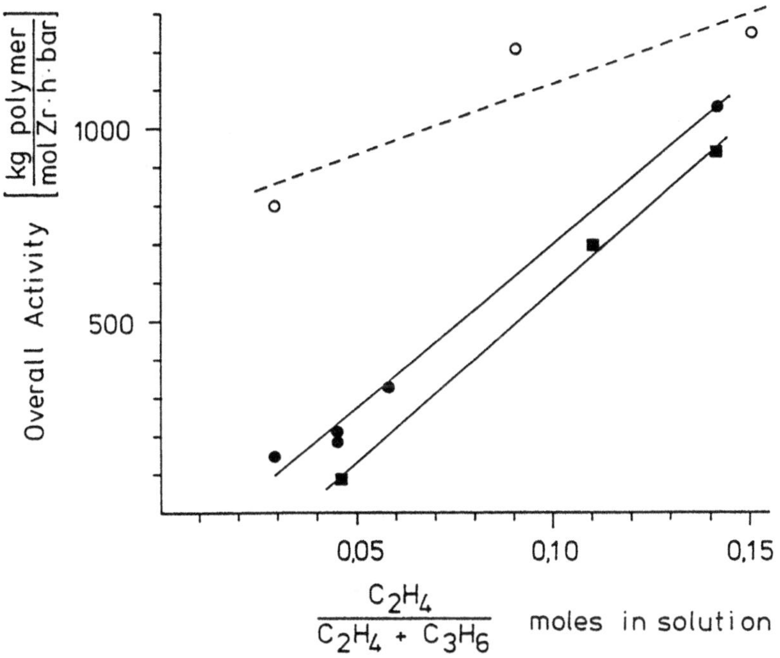

Fig. 4: Overall activity versus the monomer composition in solution.

—o— : diene-free copolymerization with (Zr) = $5 \cdot 10^{-7}$ mol/l

—●— : terpolymerization with 0.1 mol/l ENB and (Zr) = $5 \cdot 10^{-7}$ mol/l

—■— : terpolymerization with 0.1 mol/l ENB and (Zr) = $1 \cdot 10^{-6}$ mol/l

Fig. 4 shows that lower ethylene/propylene ratios decrease the activities. By comparing the terpolymerization with the diene-free copolymerization it also becomes obvious that the presence of the diene monomer causes lower activities, especially at low ethylene/propylene ratios. The polymers prepared have an ethylene to propylene ratio between 80:20 and 50:50 wt.-%. The norbornene content lies between 2-10 wt.-%, the molecular weight between 50,000 and 220,000. For the polymers $r_1 \cdot r_2 = 0{,}3$ was measured which corresponds to a random distribution of ethylene and propylene. Molecular weight distributions (M_w/M_n) given by GPC-curves have an average value of only 1.7. The properties of the obtained EPDM-elastomers are comparable to those produced with vanadium catalysts.

6. POLYMERIZATION WITH CHIRAL ZIRCONOCEN CATALYSTS

If the π-bonded cyclopentadienyl ligands of the zirconium compound are changed for ethylene-bridged indenyl-rings a stereorigid chiral-zirconium is formed (Fig. 5).

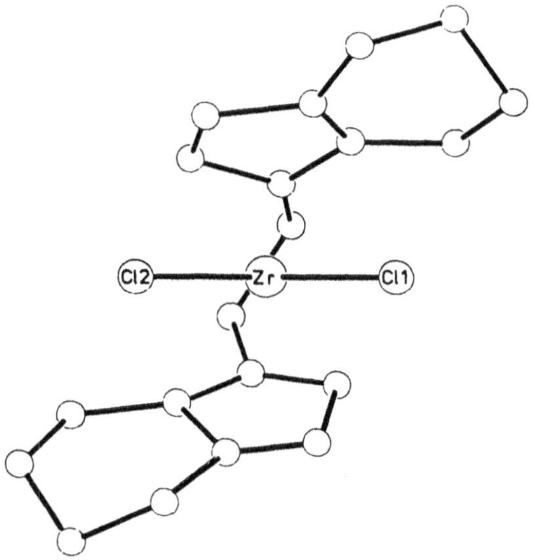

Fig. 5: Structure of rac-Ethylene(bistetrahydroindenyl)zirconiumdichloride[5,6]

Contrary to the analogous titan compound the chiral zirconocene forms solely the racemic mixture and no additional meso-form.

Using the chiral rac-ethylenebis(tetrahydroindenyl)zirconiumdichloride together with methylalumoxane as catalyst we were able to obtain pure isotactic polypropylene. Table V shows the polymerization conditions and the activity.[9]

Table V

Polymerization of 70 ml propylene with rac - Et - $(TH\text{-}Ind)_2ZrCl_2$ ($8{,}4\cdot10^{-6}$ mol/l) and 1,6.10 mol Al/l methylalumoxane in 330 ml toluene.

Polymerization temperature (°C)	Time (min)	Yield (g)	Activity (kg PP/mol Zr·h)
-20	360	1,5	80
-10	270	4,5	300
0	255	12,5	880
8	180	13,0	1 300
15	170	26,7	2 900
20	120	31,3	4 750
60	90	38,7	7 700

The activity increases with increasing polymerization temperature and reaches a value of 7,700 kg isotactic polypropylene per mole Zr and h. The properties of the obtained polypropylene are extraordinary (Table VI).

Table VI
Properties of isotactic polypropylene obtained with rac - Et $(TH\text{-}Ind)_2ZrCl_2$.

T (°C)	$\overline{M}_w$	M_w/M_n	Soluble Prod.	Isotact.Ind.
-10	305 000	2,6	0,25 %	90,0
0	144 000	2,4	0,2 %	88,1
15	62 000	2,0	0,7 %	87,3
20	45 000	1,9	1,0 %	86,0

The toluene soluble part consists of less than 0,2 weight percent and therefore lies below that of isotactic polypropylene produced with supported catalysts. The high index of isotacticity, determined via the method of Luongo, together with the frequency of triads in nmr experiments (95,9 mm; 3,2 mr; 0,9 rr) suggest that propene is exclusively head to tail bonded with the same overall configuration. Only every 50th propene unit has the wrong configuration.

Molecular weights are strongly dependent on reaction temperature as is shown in Fig. 6.

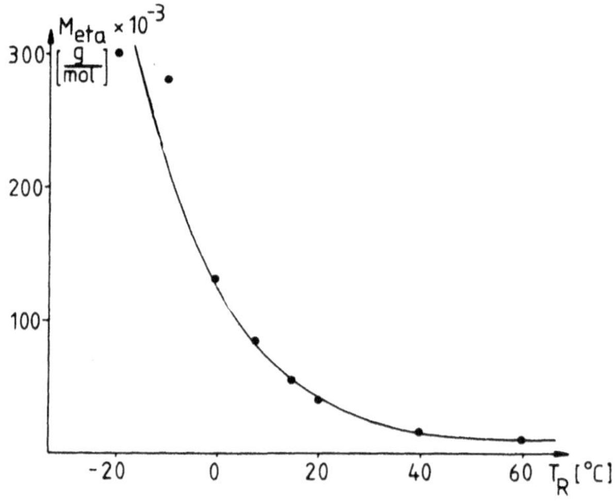

Fig. 6: Molecular weight of isotactic polypropylene in dependence of the polymerization temperature.

By variation of reaction temperature in the range from -20 °C to 60 °C molecular weights from 300,000 to 1,500 are yielded. Working at +10°C generates molecular weights between 50,000 and 200,000 which is the relevant range for the manufacturing technology.

Using one of the optically enantiomers as catalyst instead of the racemic ethylenebis(tetrahydroindenyl)zirconiumdichloride yields optically active polypropylene (Table VII).

Table VII

Synthesis of optical active polypropylene obtained with the catalyst Et(TH-Ind)$_2$ZrCl$_2$/methylalumoxane

optical rotation of the catalyst	: +280 °
Zr	: $1{,}26\cdot10^{-5}$ mol
-Al/CH$_3$)O-	: $5{,}32\cdot10^{-3}$ mol
propylene	: 65 ml
solvent	: 330 ml toluol
temperature	: 15 °C
yield/2h	: 7,5 g
optical rotation of a polypropylene film	: -0,7°(d = 0,1 mm)

A thin transparent film of 0,1 mm thickness, prepared from the highly isotactic polypropylene is able to rotate linearily polarized light about an angle of 0,7°. In the author's mind this is the first time that optically active polypropylene has been produced. The overwhelming majority of all propylene unities are inserted into the polymer with the same configuration, thus resulting in the preferred establishment of a helix of one screer sense. As an especially designed catalyst for polymerization of this kind the stable homogeneous zirconocene compound seems to be predestinated. Parallel to this finding the conventional assumption could be confirmed, that isotactic polymerizations need necessarily chiral centers[8].

It has become evident that homogeneous zirconocene or titanocene compounds in connection with methylaluminoxanes provide a wide array of up to now unknown possibilities in Ziegler-Natta-catalysis. This is due to the following advantages as high activity (based on the transition metal component), regiospecifity and stereoselectivy which clearly favour homogeneous catalysts over heterogeneous.

Literature

1. H. Sinn, W. Kaminsky, Adv.Organomet.Chem. 18, 99 (1980)

2. W. Kaminsky, M. Miri, H. Sinn, R. Woldt, Makromol.Chem.Rapid Commun. 4, 464 (1983)

3. J. Herwig, W. Kaminsky, Polymer Bull. 9, 464 (1983)

4. J.A. Ewen, J.Am.Chem.Soc. 106, 6355 (1984)

5. F.R.W.P. Wild, L. Zsolnai, G. Huttner, H.H. Brintzinger, J.Organomet.Chem. 232, 233 (1982)

6. F.R.W.P. Wild, M. Wasiucionek, G. Huttner, H.H. Brintzinger, J.Organomet.Chem.,in press

7. W. Kaminsky, Naturwissenschaften 71, 93 (1984)

8. P. Pino, R. Mülhaupt, Angew.Chem.Int.Ed.Engl. 19, 857 (1980)

9. W. Kaminsky, K. Külper, H.H. Brintzinger, F.R.W.P. Wild, Angew. Chem in press

A. Guyot and R. Spitz

CNRS CONTRIBUTION TO THE PHILLIPS CATALYSTS

ABSTRACT

The work done by the CNRS group was carried out in three successive waves. The first one (1960-1968) with 3 thesis students (J.C. Daniel, A. Revillon and M. Durrieu) was mainly devoted to the kinetics of propylene polymerization and to the dispersion state of the chromium oxide onto a silica-alumina support, studied by a combination of oxygen chemisorption and magnetic susceptibility measurements. In the second wave (1968-1974 with G. Vuillaume and R. Spitz) an in line method was used to follow both the activation process and the catalytic activity for ethylene and propylene polymerization. That study lead to propose a new model where the active centers are a set of chromium atoms on the border of the chromium oxide clusters. The last wave (1976-1983, B. Florin and G. Mabilon) was devoted to modified catalysts and copolymerization of ethylene with α-olefines, using silica as a support. Again, it was concluded that the active centers were probably not isolated chromium atoms. The paper includes a final discussion of two open problems, the mechanism for the initiation reactions, and the question of the possible distribution of active sites of various activity and properties which may explain the broadness of the molecular weight distribution.

INTRODUCTION

The evolution of the Phillips process of ethylene polymerization involves progress in the process (from the solution process to the particle form suspension process), as well as in the catalysts. For the latter, the early Phillips catalysts were mostly based on silica-alumina support. Then came the silica support, especially with controlled pore size. Finally, came silica supports

R. B. Seymour and T. Cheng (eds.), History of Polyolefins, 271–305.

modified by addition of titanium compounds and fluorosilicate compounds. The main driving forces for the change in the catalyst were the need for a better control of molecular weight and molecular weight distribution, and more recently of the comonomer efficiency for the modified polyethylene (linear low density polymers).

For a long time, the work done by the CNRS group has been dealing with the first generation of catalysts. Five thesis were completed on this line : by J.C. Daniel (1964), A. Revillon (1967), M. Durrieu (1968), G. Vuillaume (1970) and R. Spitz (1971). The work was devoted to the kinetics of both ethylene and propylene polymerization either in a batch reactor or in an in-line reactor, and also and maybe chiefly to the interaction of chromium oxide on the support and its dispersion state ; the conclusion of these studies was the proposal of a new model of sites, where the active chromium atoms were located at the boundary of clusters of chromium oxide dispersed on the support surface. One might be surprised that the polymerization of propylene was initially chosen ; although the initial Phillips patent (1) described that polymerization as carried out with a limited success, it was the basis for one of the most famous "case" of industrial property of the polymer history. So, a special paragraph will be devoted to that study in the present review.

More recently, we were interested in copolymers of ethylene and α-olefins, using modified silica supported catalysts. Two thesis were completed on that line, by B. Florin (1977) and G. Mabilon (1982) ; the latter included a new study of the catalyst which supported the conclusion that the active sites were probably not isolated chromium atoms.

In the last part of this review, we will discuss the significance of our contributions in the historical context of the field, and will point out the unsolved problems.

A - SILICA-ALUMINA BASED CATALYSTS

1 - Kinetics of propylene polymerization at medium pressure (2)

An original method for following the kinetics of propylene polymerization has been developed ; it was based on the gas chromatographic analysis of a mixture of propane and propylene (3) ; the first one serves as an internal calibration for the second. Most of the experiments were carried out between 90 and 120°C, i.e. in a temperature range where the mixture (always including a small amount of ethane) is above its critical temperature. The reaction is first order, versus monomer concentration and follows an Arrhenius law with an activation energy of 35 Kj/mole for the more active catalysts. The polymerization rate depends on pressure only with respect to monomer concentration, but, when the pressure is low and the temperature high enough, a second process occurs, which causes the formation of volatil oligomers and retardation of the normal polymerization process ; the formation of volatil oligomers is proved by the difference between the consumption of propylene and the weight of polymer obtained.

More sophisticated chromatographic analysis were used to follow both the consumption of the monomer and the formation of oligomers (4) ; it was shown that, if the polymerization activity decreases, for instance through improper activation process which leads to a low average oxidation degree of the chromium, the oligomerization activity increases. The oligomerization activity is caused by the acidic center of the silica-alumina carrier (5) ; upon dealumination by acidic treatment that activity disappears ; the oligomerization activity of the carrier is largely affected by the deposition of chromium oxide and a subsequent proper activation treatment.

The polymerization reaction is also first order versus the catalyst amount, but this law is verified only when the purification of the reactants is carried out at a high enough level ; the catalysts can be poisoned by a very low amount of various compounds such as allyl alcohol or propionaldehyde. Inhibition experiments (6) lead to conclude that the proportion of chromium atoms which are active sites is lower than 10 %.

2 - <u>Products of propylene polymerization</u>

Three kinds of products can be obtained from propylene polymerization with chromium oxide - silica alumina catalysts : volatil hydrocarbons, atactic low molecular weight (1500) polymers, semi-crystalline high polymers with molecular weight ranging from 10.000 to 100.000 or more. Alcenes are present among the volatil hydrocarbons, they can partly be copolymerized with propylene, and by this way they reduce the polymerization rate ; these low molecular weight products come from the oligomerization process and their proportion is normally from 10 to 30 %. The proportion of the atactic fraction is from 60 to 80 %, leaving 10 to 15 % of semi-crystalline polymer. Progress are made using a silica-alumina carrier containg 4 % sodium , then, the oligomerization process is partly inhibited and the amount of semi-crystalline fraction, insoluble in ethylether may reach 25 % with increasing molecular weight ; however, half of that material is soluble in boiling heptane.

3 - Kinetics of ethylene and propylene polymerization and copolymerization at low pressure

In a second part of the work, a small integral dynamic reactor was used for the continuous measurement of the instantaneous polymerization rate (7) : the catalyst is suspended in a stream of gas (activation gas, helium, ethylene or even propylene) and a differential catharometer allows the measurement of the consumption of that gas. With ethylene, the rate goes through maximum after a few minutes, then levels off, and finally decreases, owing to the blocking of the surface by the polymer (fig. 1). The maximum level allows to calculate safely the catalytic activity. The results are very sensitive to the activation process : depending on it, one may or may not observe a maximum of the catalytic activity between 90 and 120° C when the temperature is varied from 0 to 180° C. Reliable results are obtained if the last step of the activation is carried out with small amount of monomers between 150 and 250° C, Then, the reaction order versus monomer concentration is 1 at 145°C, and decreases to fractional orders down to 0.25 at -50°C (8), while a maximum rate is observed at 90°C. These features can be explained assuming a Langmuir-Hirnhelwood mechanism in which the monomer is first adsorbed at sites near the active sites ; depending on the temperature, the rate determining step may be either the adsorption or the migration and the insertion of the monomer.

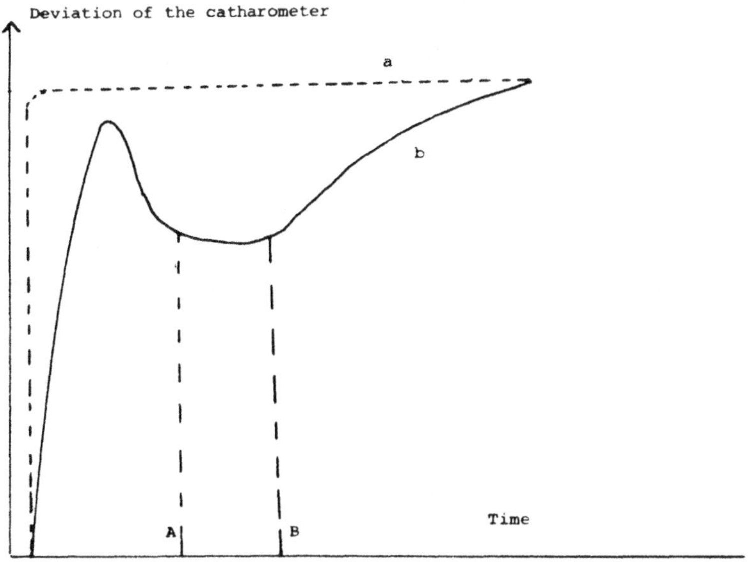

Figure 1 a/ without catalyst

b/ with catalyst : A end of the induction period

B end of the steady state conditions before progressive blocking

The same apparatus was used to study the polymerization of propylene (9). Large differences are observed bewteen ethylene and propylene : the polymerization rate is much higher for ethylene than for propylene.The ratio of the rates is the highest at high pressure and decreases by lowering the pressure from about 100 under 50 bars to about 10 at 1 bar, and less at under atmospheric pressures. The shapes of the kinetic curves are very different. A maximum rate is observed after a more or less long induction period in the case of ethylene, and in many cases, the activity remains stable, while the rate decreases continuously and rapidly in the case of propylene. The maximum activity as a function of chromium content (for the same silica-alumina support) is located at about 2 % Cr for ethylene, and between 4 % at low pressure and about 6 % at high pressure for propylene.

In both cases, a maximum of the activity as a function of the temperature is observed, but the decrease of activity by increasing the temperature is much larger in the case of propylene. However, there is no maximum of the initial activity for the propylene after a reducing pretreatment has been carried out. Whatever the conditions of activation and eventually of reducing pretreatment, the polymerization is always first order versus propylene pressure at all measurable ranges of temperature and pressure, while the order decreases from 1 to low values in the case of ethylene by decreasing the temperature at low pressure, and also during polymerization.

Further, the apparent activation energy is higher in the case of propylene (42 Kjoule instead of 0 to 25, depending on the pretreatment conditions). Some copolymerization experiments were carried out : whatever the ratio of ethylene to propylene in the monomer feed is between 1 and 5, the shape of the kinetic curve is the same as for propylene polymerization ; furthermore, the polymerization rate is not higher than the homopolymerization rate of propylene, when the feed contains more that 20 mole % of propylene.

4 - Interaction of chromium oxide and the support

The silica-alumina used (Ketjen, with 13.5 % alumina) is not at all an homogeneous solid ; it is considered as a mixture of alumina spheres and silica spheres with a small amount of co-oxide zones, which are viewed as elongated domains of very small size (10). These domains are sensitive to acidic attack, and, by that way, more or less complete dealumination of the support may be carried out. The preparation of the Phillips catalysis involves first the impregnation of the support with an aqueous solution of chromic acid, and then the activation by thermal treatment under a stream of air which causes the dehydration of the catalyst, and leads the chromium atoms to a high average oxidation number.

Interaction between chromic acid and the support already takes place during the impregnation (11). After it has been completed, a small part of both silica and alumina became extractible by water ; a small drop in the surface area is observed (50 m^2/g from 620 m^2/g). Differential thermal analysis carried out during the heating treatment shows in addition to an important endothermic effect between 100 and 300°C due to dehydration, an exotherm between 250 and 420°C. This exotherm is mainly sensitive to the amount cf alumina of the support ; partial dealumination of the support through HCl treatment before impregnation causes a large drop in exotherm area ; there is no exothermic peak if the support is silica. It was shown that a partial reduction of the chromium oxide is to be associated to that exotherm , that reduction involves about 0.3 g chromium per 100 gram catalyst, which further cannot be reoxidized during the following activation treatment, even if it is carried out under oxygen.

Finally, thermal treatment of the catalyst causes a large drop in the surface area of the catalyst (12). Starting from a support with 620 m^2/g, one gets a catalyst with a surface area between 340 and 465 m^2/g, depending on the chromium contents, its final average oxidation number, and the temperature of thermal treatment. Except for the very high temperature, where the surface area of the support itself may be affected, that drop is also to be associated with the interaction between chromium oxide and the support; in addition, all the factors that cause more reduction of the chromium oxide (higher chromium contents, higher activation temperature, lower oxygen contents of the activation gas, lower spatial velocity of the stream of gas), also causes a larger drop in the surface area (figure 2).

Oxygen chemisorption measurements (12), carried out after reduction by hydrogen at moderate temperatures, were used to estimate the area covered with chromium.For a typical catalyst (support with 13.5 % alumina and 3 % chromium contents) that area is around 15 m^2/g after impregnation, but before activation, and increases up to 23 m^2/g after proper activation. For a fixed amount of chromium, the area increases with the average oxidation number. When the amount of chromium is varied, the area covered with it goes through a maximum of 27 m^2/g at a chromium contents of 7 % and then slowly decreases (20m^2/g for 12 % Cr). After extraction of the hexavalent chromium with water, the area covered with trivalent chromium may be studied, it is very low (about 2 or 3 m^2/g) for a well activated catalyst. If the support includes sodium, the chromium is more clustered. On the other hand, it is more spread, if the support is silica (Table I).

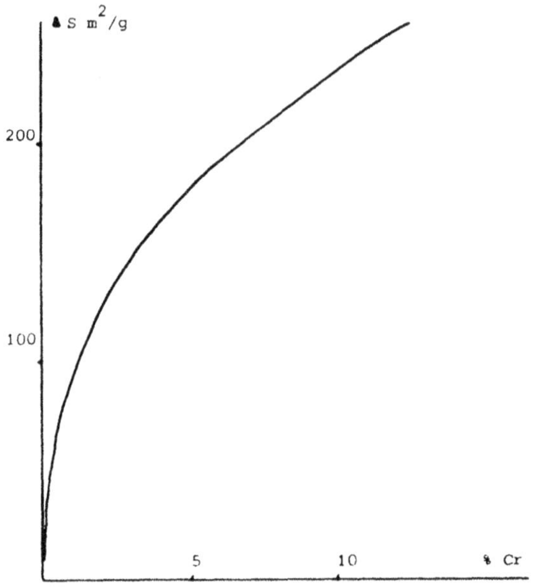

Figure 2 :

Reduction fo surface area in Phillips catalyst supported on silica-alumina after activation

Attempts were made to characterize the size and shape of the chromium oxide domain ; at low temperatures, the superantiferromagnetism of the dispersed trivalent chromium oxide can be observed ; the material has two sublattices of ions with antiparallel spins and, when its size is low enough, the contribution of the edges makes the mutual cancellation of the two sublattices incomplete, so that paramagnetism is measured. From these measurements, the interpretation, according to Neel's theory (13), suggests that the hexavalent chromium tends to agglomerate on the trivalent chromium which is located in anisotropic domains, which correspond most probably to the cooxide zones of the support (14).

From these studies, some correlation between the activity and the average number of chromium oxide layers in the chromium oxide domains may be drawn, showing a maximum for about 2 chromium oxide layers (12) (Figure 3).

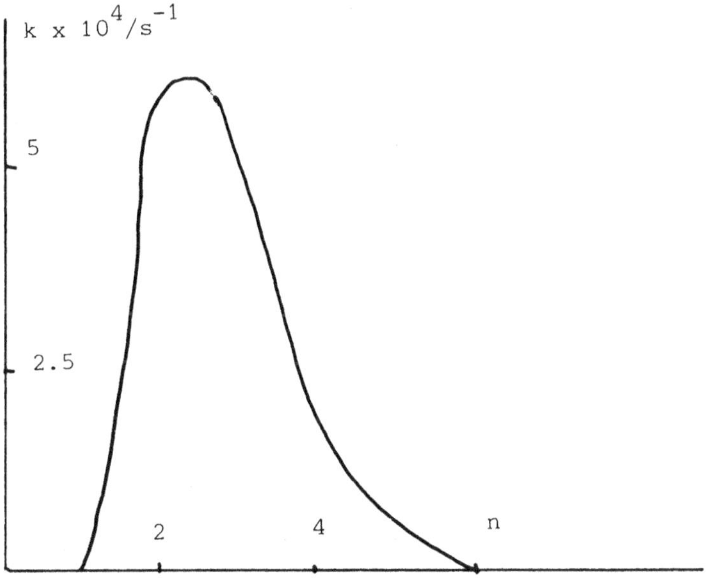

Figure 3

Catalytic activity (k) versus the number n of chromium oxide layers

5 - Activation of the catalyst

Although the term activation usually refers to the thermal treatment of the catalyst under a stream of gas, it was shown that the last step of the catalyst treatment is actually a reduction by the monomer itself prior to polymerization. The reduction can be carried out by other compounds, CO being the most often mentioned (15). We have studied these two steps independently (16). For the first step, a set of parameters have been studied : the activation temperature (between 300 and 700°C), the heating rate to reach that temperature, the duration of the treatment at that temperature, the gas nature, the velocity of the gas stream. These parameters may influence the average oxidation number of the chromium, its dispersion state and finally the activity versus olefin polymerization (Table I).

As already stated, the interaction between the support and the chromium oxide is rapid and takes place mostly during the heating process, before the nominal temperature has been reached ; it causes a partial reduction of the chromium to the trivalent state, in domains in which more chromium tends to agglomerate. The average oxidation number is decreased by increasing the nominal temperature or decreasing the velocity of the gas stream and its contents in oxygen. At high temperatures, the hexavalent chromium oxide seems to be a little bit volatil, then, upon increasing the treatment duration, the hexavalent chromium oxide tends to spread onto the surface, in more anisotropic domains, while the trivalent chromium tends to agglomerate in more isotropic domains (17). Also, at very high temperature, the surface area covered by the chromium tends to be higher.

In ethylene polymerization, the activity is not governed by the average oxidation number, provided it is higher than 4.5 ; this is at variance from the propylene polymerization, in which a maximum is observed for a value of 5.6. The maximum number of sites depends mostly on the temperature of the activation process ; the process duration is not important as soon as it is long enough to cause a complete dehydration of the solid ; the changes observed in the spreading of the chromium oxide are not related to any change in the activity.

Different authors (15, 18-22) have shown that the activity was dependent on pretreatment of the catalyst with various chemicals such as CO, H_2 or limited amounts of the monomer or other hydrocarbons. We did a systematic study using H_2 and also ethylene. In the case of H_2, we found, for the activity for ethylene polymerization, a very sharp maximum (figure 4), depending on the duration of the pretreatment at a given temperature, as well as on the chromium contents.

Table I - Surface area and activity versus propylene polymerization for various catalysts

Support % alumina	Cr contents W/W %	Activation under gas	VSH	temperature °C	Surface area (m^2/g) Catalyst	Cr	Activity $x10^5/s^{-1}/g$
0.6	3	Air	400	550	550	27.8	3
7.4	2.85	Air	400	550	490	26.7	9
13.5	3	Air	400	550	425	18.1	10
24.8	3.17	Air	400	550	374	14.9	4
13.5[a]	1.47	Air	400	550	344	5.1	1.2
0.015	2.35	Air	400	550	435	27.2	1.5
0.015	2.35	Air	400	700	366	-	6
13.5	0.51	Air	400	550	545	4.1	0.5
13.5	1.06	Air	400	550	514	8.3	2.5
13.5	2.9	Air	400	550	415	14	6
13.5	5.0	Air	400	550	384	27.2	16
13.5	6.84	Air	400	550	313	21.8	5.2
13.5	8.97	Air	400	550	256	20.4	0.1
13.5	12.65	Air	400	550	256	20.4	0.1
13.5	3	N_2	400	550	-	5.7	0
13.5	3	Air	50	550	-	9.8	35
13.5	3	Air	200	550	-	18.1	10
13.5	3	Air	400	550	-	22.7	14
13.5	3	O_2	400	550	-	22	13

(a) : sodium contents : 4 %

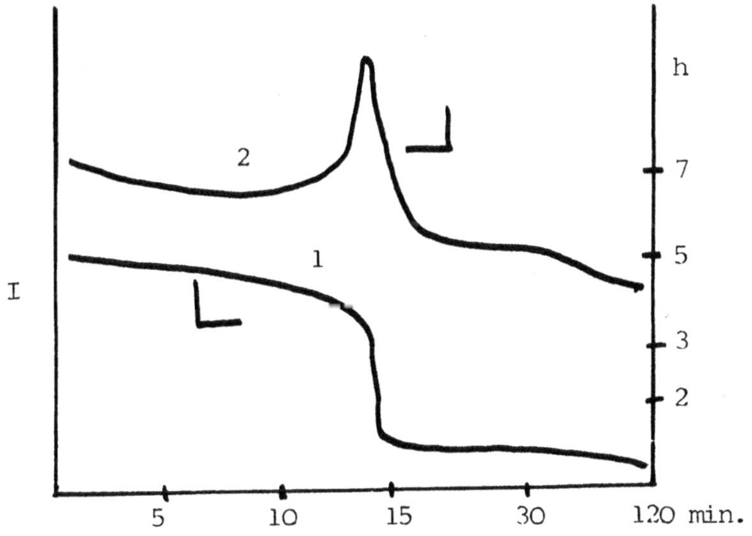

Figure 4 - 1 intensity (arbitrary units)

2 ethylene activity

Of course, the temperature of the pretreatment is another important parameter , it must be carried out between 250 and 300 °C, temperatures at which the water coming from the chromium oxide reduction is easily desorbed. The pretreatment with H_2 causes the induction period before reaching the maximum polymerization rate to decrease.

The induction period is also very much decreased if the pretreatment is carried out by the monomer itself : the pretreatment may be carried out using a limited amount of monomer (no more than 4 mmole per g of catalyst) at a temperature higher than the polymerization temperature. It causes the reduction of the catalyst and the formation of oxidized products, which must be desorbed (eventually by the monomer itself). The reducing power of ethylene is very dependent on the temperature ; at 250 °C, the optimum pretreatment is very short, otherwise, the catalyst is too much reduced and the activity drops. Propylene and other hydrocarbons can be used, but their reducing power is much stronger, so that the optimum involves shorter contact times; with propane, the activity that can be reached is lower, probably because oxidized products cannot be easily desorbed. It is important to note that, after the optimum pretreatment, the catalyst activity is stabilized against further reduction by the monomer at the polymerization temperature although it is clear that the last activation step is a reduction of the active chromium atom, probably to the oxidation number of 2 (25).

As a conclusion, the activation of the supported chromium oxide catalyst involves five essential steps before the polymerization process may begin :

1 - interaction between chromium oxide and support. This is essentially a rapid process which depends on the temperature and which fixes the dispersion state of the chromium and the number of potentially active sites.

2 - Dehydration and desorption of water. This is a time- and temperature-dependent process.

3 - Reduction of the chromium oxide to a definite state. This depends on the temperature and the amount of reducing agent ; it requires well-defined conditions to obtain a catalyst surface with high activity.

4 - Desorption of the oxidized products. This depends on the temperature and on the nature of the reactants.

5 - Complexation of the monomer onto the sites.

An important difference appears between the results of treatments with H_2 and C_2H_4. With the hydrocarbon, the site is actually active and stabilized against impurities. This is shown by the results of low temperature (0^oC) polymerization of C_2H_4 (4, 8) where an activity is found only after pretreatment with C_2H_4 or CO.

After a conventional activation process, the easiest way to reach the maximum activity is by pretreatment with a properly limited amount of ethylene at a rather high temperature.

6 - Model of the active sites

Starting from the fact that a small part of the chromium atom is actually active for polymerization, the chromium oxide tends to agglomerate on clusters located on specific cooxide zones of the silica-alumina support, the activity versus ethylene polymerization goes through a maximum for 2-3 % chromium and reaches its maximum value after a proper reducing treatment which leads the average oxidation number of the chromium to be around 4, we have elaborated the following models for the active sites (24) : they are formed with a set of three chromium atoms Cr^{+3}, Cr^{+6}, Cr^{+3}, located on the boundary of the chromium oxide domains distributed onto the surface of the support. This model is valid only for the silica-alumina support and also mainly accounts for catalysts with more than 1 % chromium.

According to that model, the activity may be estimated as being proportional to the perimeter of the chromium oxide domains ; assuming, for the sake of simplicity, that they are spherical, their radius is then proportional to the ratio between the area covered by the chromium and the activity. The curve showing the dependence of the radius of these chromium domains on the chromium contents is shown in figure 5. The second curve in the same figure shows the time necessary to get the maximum activity for ethylene polymerization for a pretreatment with H_2 at 295°C.

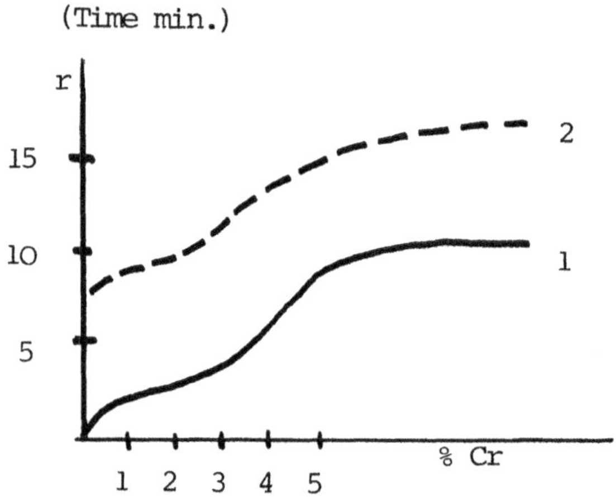

Figure 5 - 1 radius of the chromium oxide side

2 time length for optimum activity of H_2 pretreatments

The parallelism between these two curves is striking and suggests that the pretreatment with H_2, which leads to an average oxidation number of 4, corresponds to the formation of a maximum number of sets of Cr^{+3}-Cr^{+6}-Cr^{+3} vicinal chromium atoms ; this number rapidly changes with time at the rather high pretreatment temperature and for that reason goes through a very sharp maximum for the optimal treatment.

The model also involves that the monomer is first adsorbed onto the chromium domains, and then jump from site to site up to the active sites located at the boundary ; there, the polymer is formed and is growing on the silica side of the support while progressively blocking the active sites. The data supporting these statements are :

* i - **the fact that the surface area of the whole solid drops vary soon more rapidly than the activity which decreases according to an exponential law** and,

* ii - **the fact that during polymerization the reaction order decreases, what means, in the Langmuir-Hinshelwood mechanism, that the surface becomes more and more saturated with the monomer ; the latter fact tends to show that the chromium oxide domain remains free of polymer and then available for adsorption of the monomer.**

The correspondence between the polymerization activity and the intensity of the ESR spectrum narrow line, associated to the so-called γ-phase, was studied too (26). Such a correspondence, initially pointed out by Kazanskii and Turkevitch (19) who assigned the signal to Cr^{+5} ions, has been the subject of many discussions. It is qualitative but not quantitative. The best explanation seems to consider that the -phase should be associated to a set of three vicinal chromium atoms Cr^{+6}-Cr^{+3}-Cr^{+6} which have of course an average oxidation number of 5. This agrees with an hypothesis, early suggested by Ellison and Singh (27). A strong argument is the change in the γ-phase signal which is observed upon pretreatment with H_2 (figure 4).

The proposed model also allows to explain the difference in the catalyst behaviour versus ethylene and propylene polymerization, taking into account the differences between these two monomers in their properties of reduction of the chromium oxide, and of adsorption onto the catalyst surface (28). In the absence of a reducing pretreatment with H_2 or C_2H_4, the reduction by propylene is very rapid and the chances to get a large number of active sets is the highest with the highest initial average oxidation number of chromium ; the induction period is not observed because reduction is rapid enough, so that a large number of sites are immediately formed (reduction of the central atom to

Cr^{+2}). If a pretreatment with H_2 was carried out under optimal conditions, many sites are stabilized, and their deactivation by further reduction is less easy ; then, at a sufficiently low temperature (20 to 60°C), the shape of the kinetic curve for propylene polymerization is the same as for ethylene, i.e. with an induction period.

After optimal reducing pretreatment, the ethylene polymerization rate is practically independent on the temperature in the 0-150°C range ; the adsorption of ethylene upon Cr_2O_3 (which is not a polymerization catalyst) is also independent on temperature. On the other hand, an activation energy of about 40 Kj is observed for propylene polymerization ; this is thought to be due to the dependence of propylene adsorption with temperature, because in all the range the order versus monomer pressure is 1, while it is between 0 and 1 for ethylene ; so, the ethylene adsorbed on the inner part of the chromium oxide domains is much more mobile than propylene adsorbed at the same place.

It remains to explain why, in copolymerization experiments, the kinetic behavior is that of propylene. Sequential experiments (28) have shown that, if after an optimal pretreatment with ethylene, propylene polymerization is carried out for a limited time at moderate temperature, and then ethylene polymerization, a very long induction period is observed, but the final activity is not decreased, as compared with the activity of a regular ethylene polymerization experiment carried out after the same optimal pretreatment. It means that the active sites have not been destroyed during the propylene polymerization step, but the adsorption of propylene is very strong. A rational explanation of these facts is suggested by the experiments described by Hogan(29) and Krauss et al. (30), who have shown independently that, after activation pretreatment with CO, the active sites may form complexes with N_2,CO or ethylene. These complexes probably involve two molecules of gas ; further mixed complexes with CO and ethylene were not active in polymerization.These facts imply that the active sites may be blocked by strong CO adsorption on one of the two coordination positions available for complexation in the site. It may also imply that the development of polymerization activity needs the adsorption of two monomer molecules onto the site. If such a situation is valid for all the active sites, it may be suggested that one molecule of propylene adsorbed on a site may inhibit ethylene polymerization up to the moment when that molecule is either desorbed or polymerized.

B - SILICA AND MODIFIED SUPPORTED CATALYSTS

1 - Interaction and activation

The interaction between chromium oxide and silica is much less specific than when the support is silica-alumina : during the impregnation and activation there is no interaction strong enough to lead to a DTA exotherm, or to cause any noticeable change in the surface area. According to Hogan (29) and also Mc Daniel (31), chromic acid does react with the surface silanol groups of the silica to give mostly monochromates and eventually dichromates. However, infrared data of Zechinna et al (32) point out the formation of di and polychromate. Our infrared data are in agreement with those of the italian group. This is not surprising because the structure of CrO_3 which is a polymer in the solid state, remains partly associated in the concentrated solutions used for impregnation (33).

A few sets of catalysts with chromium contents between 0.4 and 4 % were prepared after impregnation of a silica (GRACE 952) with an aqueous solution of CrO_3. The concentration of the solution (between 0.25 and 2.5 N) is calculated so that its volume (0.3 g/g catalyst) corresponds to about 1/5 of the porous volume of the support. The powdered impregnated catalyst is then slowly dried under argon up to 200°C. The activation is carried out under air (or mixture of oxygen (23 %) and argon) at a temperature varying from 600 to 750°C. The catalyst is cooled down under argon under 300°C to room temperature. In such conditions practically all the Cr is hexavalent and watersoluble.

Reduction profiles have been obtained as previously done either by Holm (23) or Hogan (29), using either H_2 or CO. In the case of H_2, the experimental conditions were chosen so that to limit the amount of water present in the system (stream of H_2 in argon). A maximum in the H_2 consumption (as measured with a catharometer) is observed at temperature decreasing from 400°C (for 0.4 % Cr) to about 350°C (for 4 %). In the last case a double wave is visible. The profile is very broad between 200 and 530, which shows that even if it is very diluted the chromium may be agglomerated in clusters easy to reduce. From the amount of H_2 consumed most of the Cr is reduced to Cr^{III} but a part of it is more reduced. The reduced catalyst is easily reoxidized under O_2. On the other hand, a catalyst treated at 750°C under argon is not reduced further by H_2. In the case of CO, the Boudouard reaction (2 CO $\rightleftarrows$ CO_2 + C) give a carbonaceous residue deposited when the temperature is higher than 450°C. For that reason the temperature has been limited under 430°C (0.1 bar of CO). Only one maximum is observed at temperature decreasing from 360 to 280°C when the Cr contents increases from 0.4 to 10 %. From the amount of CO consumed the reduction goes to the CrII state, with some agglomeration above 4 % of Cr.

Chemisorption of CO and O_2 has been also studied. In the case of CO the catalyst was first reduced by CO at high temperature and the chemisorption measured at 0°C. The amount of chemisorbed CO then corresponds to a ratio CO/Cr always less than 1 (from 0.06 to 0.49 when the Cr contents decreases from 12 to 2 %) A part of it can be desorbed under argon at 0°C ; the quasi totality of the remaining part is desorbed at higher temperature ; a maximum rate of thermodesorption is observed at temperature between 60 and 90°C, the higher temperature corresponding to a chromium contents of 2 to 4 %. In the case of O_2, the catalyst is first reduced by H_2 at 600°C, and then O_2 is chemisorbed at various temperature (between 0 and 300°C). A second reduction with H_2 is used for titration of the chemisorbed O_2. The results show that at temperatures less than 100°C, chemisorption takes place without any reoxydation of the chromium. The reverse is true at 300°C and above with formation of hexavalent Cr. The amount of O_2 chemisorbed corresponds to O/Cr ratios lower than 1, decreasing from 0.76 (0.4 % Cr) to 0.17 (4 %). In the latter case the value may be 0.37 if the reduction is carried out so that the amount of water on the catalyst is limited as much as possible. The reduction profile shows two peaks (maxima at 290 and 410 °C) whatever is the chromium contents. Finally, from the amount of chemisorbed O_2, the area covered with chromium can be estimated ; on the same bases than those used with silica-alumina support, that area is larger ; values up to 40 m^2/g can be obtained (instead of a maximum of 27 m^2/g) ; however these absolute values are quetionnable because the reduction profile of the reference compound (Cr_2O_3) is very different from that of the catalyst (it is narrower and located at lower temperature : maximum at 220 °C). Anyway, the chromium is much more spreaded onto the surface of silica than onto silica-alumina.

In that part of the work, the activity versus ethylene polymerization was not measured directly because the strong exothermicity of the reaction make difficult safe comparisons between highly active catalysts ; the method adopted was to measure the temperature at which the polymerization starts, for catalyst preactivated or not by reduction with CO. A partial reduction is carried out at various levels, then the temperature is set at 0 °C (so that physisorption of ethylene is not too important); ethylene is then introduced and the temperature slowly raised up to the point where an exothermic effect is noticed. With increasing reduction level, the starting temperature decreases from 163 °C (1 % Cr) to 0 or less but becomes positive again if the proportion of reduced Cr increases near 1. So the isolated chromium atoms, which are the more difficult to be reduced, are not the more active sites. On the contrary, as already stated by Eley (34), these more active sites are those showing intermediate reductibility. If no reduction treatment is carried out, the polymerization starts only at higher temperature between 120 (for 4 % Cr) and 187 °C (for 0.4 % Cr). This results is easily explained because the reductibility of the chromium by ethylene is easier if the chromium is more agglomerated.

2 - Effect of Titanium and Fluorine

A set of recent studies point out the interest of adding titanium compounds mainly (35-36) and sometimes fluorine compounds (37), in the catalyst formulation to improve the catalytic activity and, chiefly, to increase the value of the melt index of the polymer and to control the molecular weight distribution. Various different protocoles have been used to introduce these additives. In our studies, they have been introduced after the impregnation with CrO_3 and the drying step, but before the activation with air. The titanium was $Ti(OBu)_4$ in heptane solution, while the fluorine was finely ground $(NH_4)_2SiF_6$. During the activation, the later compound is decomposed leading to SiF_4, NH_3 and HF which may react with both Si-O-Si and SiOH. These interactions leads to a decrease of the surface area, which is the stronger for the higher amount of F, but is never so drastic, as reported by Mc Daniel (31). The presence of the titanium also causes a drop of the surface area up to about 40 m^2/g (from an initial value of 320 m^2/g) which is smaller than when both Ti and F are present (then surface area as small as 240 m^2/g have been measured). The titanium butoxide is strongly adsorbed onto the surface of the catalyst in amount which is larger than for a stoechiometric reaction with the silanol groups ; the interaction cannot be detected with DTA ; however for a large amount of Ti (10 %) some volatilization of the titanate takes place with combustion which give some CO and CO_2.

The following changes caused by Ti and F are observed in the reduction profile : in the case of reduction with H_2, the maximum is observed at lower temperatures, but the profile is broader, chiefly for the higher amounts of Ti. The effect of F is about the same but only if the amount of F is very high (more than 3 %) ; the presence of F reduces the effect of Ti. The titanium obviously causes some agglomeration of the chromium so that the amount of CO used for its reduction is reduced. This effect of agglomeration is observed also in the oxygen chemisorption experiments ; the high temperature peak for the reduction by H_2 of chemisorbed O_2 is reduced, and the ratio O/Cr also (from 0.57 to 0.41 for a catalyst with 1 % Cr and 0 to 5 % Ti) ; thus the chromium is not covered with Ti but only more agglomerated. For chemisorption of CO, the presence of titanium cause a destabilization of the chemisorbed CO ; most of it can be desorbed by argon at 0°C, chiefly if the amount of Ti is large.

In ethylene polymerization experiments, the presence of titanium causes a decrease of the minimum polymerization temperature ; this result is in agreement with the reported effect of enhancing the polymerization activity.

3 - Copolymerization experiments

Two sets of copolymers have been prepared ; in a first study (38) ethylene-hexene copolymers with quasi constant composition have been produced in liquid suspension in hexene under constant pressure of ethylene between 2 and 23 bars at temperature between 29 and 90°C, using a 2 % chromium catalyst supported on silica (Degussa - 217 m^3/g). A small amount of $Al(C_2H_5)_3$ has been found necessary as impurity scavengers ; however the catalytic activity showed a sharp maximum with the amount of scavenger ; the homopolymer produced (then hexene is replaced with hexane) is very similar to any Phillips polyethylene (one terminal unsaturation per molecule, very broad molecular weight distribution). The copolymers were observed to be composed of two very dissimilar compound : a waxy fraction with high hexene units contents (up to 89 %), and a solid fraction (with no more than 4 % hexene units) ; the amount of the waxy fraction, and also its hexene contents, increases with increasing polymerization temperature; both fractions are highly polydispersed. The hexene incorporationis decreasing very much if the amount of alkylaluminium scavenger is higher than 1 mmole per g of catalyst. A ^{13}C NMR study (39) of the copolymers shows that by increasing the amount of $Al(C_2H_5)_3$, above that limit, the proportion of hexene as isolated unit increases up to 45 %. For a limited amount of Al compound, by increasing the polymerization temperature all the sequences of hexene longer than 2 tends to increase while the amount of isolated hexene units decreases from about 30 to 20 % in the range studied. Further studies of the solid fraction (40) show that it is very similar to high density polyethylene, with a melting point at 397°K.

The second study dealt with ethylene-propylene copolymerization in gas phase, using various catalysts supported onto silica (Grace 952-320 m^2/g), the medium being fluidized by a vertical oscillating of the reactor itself. A composition drift is observed if the reactor is feeded with a constant composition monomer mixture ; the drift lead to an asymptotic equilibrium composition. A proper initial excess of propylene can be introduced so that the equilibrium composition corresponds to that of the monomer mixture continuously introduced ; then a constant composition copolymer can be produced and checked through a gas chromatographic analysis of the composition of a small amount of gas allowed to leak continuously. The copolymerization kinetics are often analysed in terms of the the usual reactivity ratios, as for the radical polymerization ; however for copolymers produced with coordination catalysts (such as Phillips or Ziegler-Natta catalysts), such an analysis which involves only one kind of active species is not correct. With these catalysts there is probably a distribution of sites with different reactivities. So it is better to speak only of relative reactivities. Assuming the reaction to be first order versus the monomer concentration, the ratio k3/k2 of the rate constant k for propylene and ethylene polymerization may be easily calculated (41). Typical results are reported on Table II. The relative reactivities are practically not dependent on the polymerization temperature and on the amount of chromium in the simplest Phillips catalysts ; however the presence of titanium greatly enhances the incorporation of the α-olefin; the introduction of fluorine only shows a similar but more limited trend,and there is no synergistic effect.

Table II - ETHYLENE-PROPYLENE COPOLYMERIZATIONS

Catalyst	Pressure (bars)	Temperature °C	Mole% propene in the feed	k3/k2
1%Cr	9.65	82	7.6	0.54
1%Cr	9.6	83	3.85	0.49
1%Cr	5	80	3.8	0.49
1%Cr	5	89	3.85	0.52
1%Cr 0.7%F	5	76	3.6	0.58
1%Cr 5%Ti	5	80	3.6	0.58
1%Cr 5%Ti 0.7%F	5	82	3.4	0.72
1%Cr 5%Ti 0.7%F	5	82	3.9	0.65
4%Cr	5	82	3.4	0.46

C - A FEW OPEN PROBLEMS

In this section, we would like to discuss the two main unsolved probems in the Phillips catalysis ; the first one deals with the mechanism for the initiation of polymerization, more precisely how the first chromium-carbon bond is formed. The second problem is the nature of the active sites, and more precisely is it only one kind of active chromium atoms or does it exist a distribution of various active species ?

Mechanism of the initiation reaction

It is well established that polyethylene produced using Phillips catalyst are very linear molecules with one saturated chain end (methyl group) and one unsaturated chain end, including a vinyl group or, less frequently, a vinylidene group. On the other hand, the number of active centers being rather small, and the productivity very high, each active center is probably producing many different molecules so that most of them are terminated and initiated through a transfer mechanism ; both spontaneous transfer (beta abstraction of an H atom and formation of a chromium hydride) and transfer onto the monomer easily explain the structure of the chain ends. The problem to be discussed here is the formation of the first polymer molecule from a given active center ; the propagation reaction obviously results from the insertion of a monomer moleculer in a chromium-carbon bond (which has been shown to exist by Krauss (54), and the termination reaction is expected to occur through a transfer mechanism, but the unsolved question is how the first chromium-carbon bond is produced. In a recent book (42), Yermakov et al stated that "there are no experimental data on the mechanism of this step". Obviously, it must result from the interaction of the monomer with the activated catalyst, because the monomer is the only one source of carbon. However, there are a few hypotheses which have been proposed in the literature. One of the oldest, assuming a tetracoordinated pentavalent active chromium was given by Van Reijen and Cossee (43) and involves the abstraction of a proton from the monomer to form an hydroxyl group according to the following scheme :

But, it is difficult to justify correctly such an hydrogen transfer. Later on, when the idea of a divalent active chromium became popular, the Russian workers (44) suggested the following scheme :

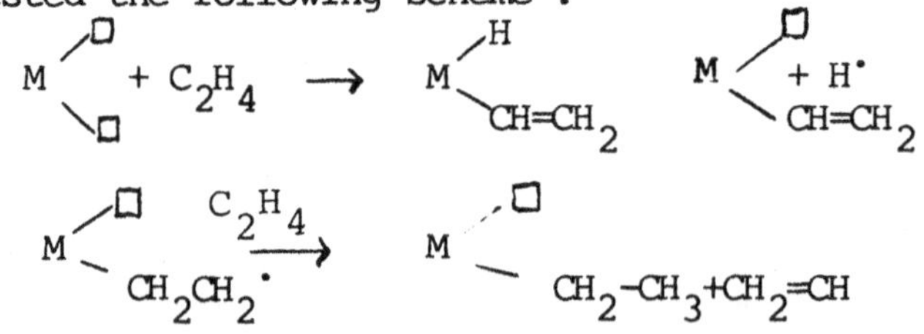

They stated, in addition, that the radicals are also able to alkylate the ions of the transition metal

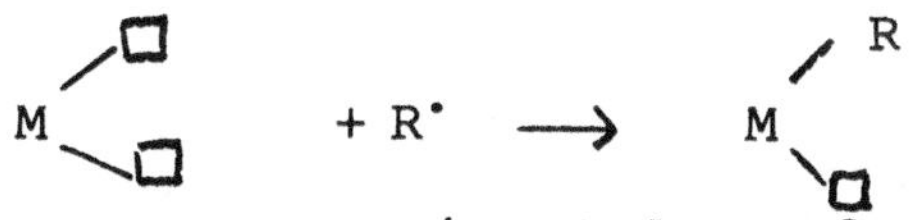

However, no experimental proof was given, and again it remains to explain the behavior of the H radical.

In the same period, allyl chromium compounds were shown to be active in ethylene polymerization (45) so that, for the Phillips catalys, an intermediate π-allyl compound was proposed, after the coordination of two ethylene molecules.

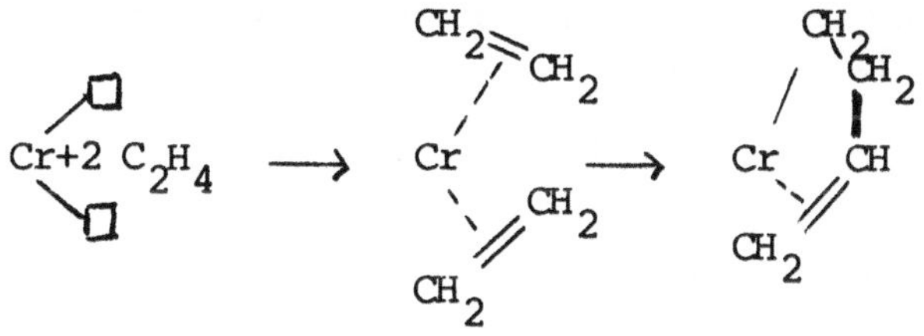

but again, such a mechanism involves the splitting of a hydrogen atom, which causes problems, and further it was shown (46) that, when π-allyl chromium compounds were used as catalyst, the allyl ligands are retained during the polymerization.

To overcome the problem of the extra H-atom, a Swedish team (47) starting from the hypothesis that active centers come from a dichromate compound, suggested that a first ethylene molecule should be inserted between two chromium atoms, previously reduced at the proper oxidation level (either II or III (48))

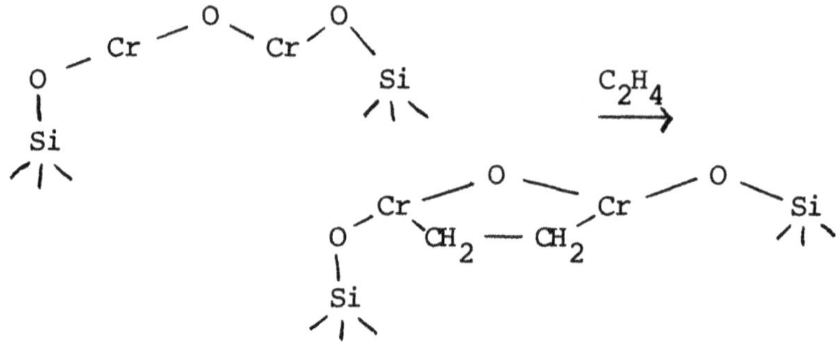

Then propagation takes place by insertion of more ethylene molecules on both sides of the complex, the termination being not considered by the authors.
The most interesting proposal was given by Krauss (49) who suggest that chromium carbene intermediate were formed. The scheme is the following :

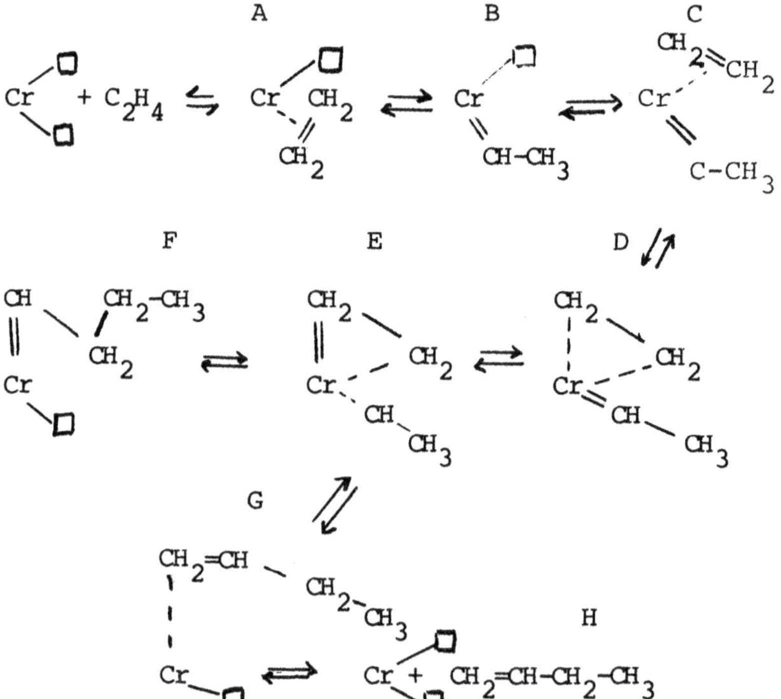

In that scheme, all the steps of polymerization, initiation (A-E) propagation and termination (G, H) involve carbene and metallocyclo compounds intermediate.

From F, steps similar to C, D and E will lead to the propagation. At moderate temperatures, Krauss observed, in addition to the high polymer, the formation of oligomeric organometallic compounds with even numbers of carbon atoms. At higher temperatures, oligomer with odd number of carbon atoms were also found ; they might be explained through a metathesis reaction :

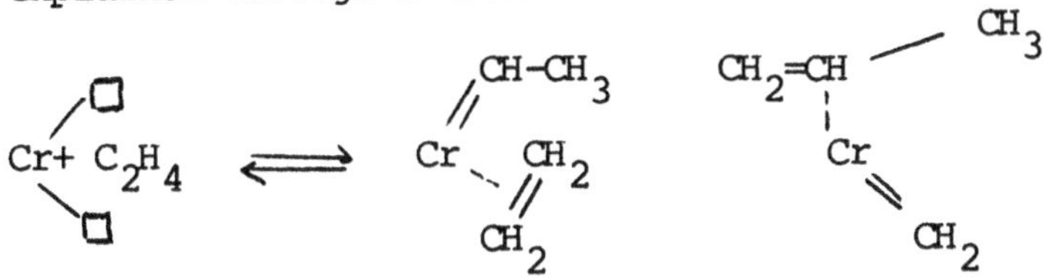

Carbene mechanisms have been well established in the metathesis reactions (50). They have also been suggested for Ziegler-Natta catalysis (51). The proof of ethylene polymerization by carbene compounds has been given by Schrock (53) ; using initially a neopentylidene hydride tantalum complex, he gets oligomers with odd number of carbon atoms including the neopentylidene residue, and others with even number of carbon atoms which are polyethylene molecules produced after a chain transfer step.

The occurence of carbene complexes in chromium based catalyst reactions is not surprising owing to the fact that chromium, molybdenium and tungsten are in the same group of the classification, the last two metals being the basis for the most active metathesis catalysts ; it can be argued also that the first stable carbene compound isolated was a chromium compound (55).

In our laboratory (56), experiments were carried out using chromium on silica activated catalysts without prereduction with CO or other compounds ; ethylene was circulated upon the catalyst at temperatures increasing from 0 °C and the starting polymerization was noticed (through its exothermic effect) at temperatures between 85 and 160 °C depending on the chromium contents ; the gas chromatographic analysis of the gas just passed onto the catalyst at the starting polymerization temperature shows the presence of a small amount of butene-1 (about 0.3 C_4H_8 per Cr atom for a catalyst with 0.4 % Cr) with no other gas. This result is in good agreement with the data reported by Krauss (57) stating that butene-1 is the main

product of the reaction of ethylene with CO reduced catalyst at 0 °C. Most probably in our case, the catalyst has been reduced by ethylene itself before producing butene-1 and then polyethylene. However, the mechanism proposed by Krauss does not explain why the production of butene seems to be a transient step just before the actual initiation of the polymerization.
The last proposal for the initiation mechanism was given by Groeneveld et al (58). They suggest that a Cr^{+3} hydride was formed from the reaction between a Cr^{+2} precursor and a OH^- ion from a vicinal silanol group. Then, the insertion of an ethylene molecule in the Cr-H bond will give the first Cr-C bond. Their hypothesis is based on in situ infrared measurements showing, at the beginning of the polymerization, the presence of a CH_3 band at 2960 cm^{-1} ; that band is not observed if, prior to the contact with ethylene, the catalyst was treated with deuterium ; according to the authors, that treatment causes the deuteration of the silanol groups, and the appearance of C-D band in the infrared spectrum of the nascent polymer. However, a counter-argument is given by Mc Daniel (62) who stated that the most active catalyst was obtained after complete elimination of the OH group upon treatment with $TiCl_4$, prior to impregnation with chromium compound under anhydrous conditions.

The distribution of active sites

It is well established that the Phillips catalysts produce polyethylene with a very broad molecular weight distribution, the main problem of that process being the control of the molecular weight. Broad molecular weight distribution can be explained on the basis of two kinds of phenomena ; either the polymerization is controlled by the diffusion of the monomer through the polymer covering the catalyst sites, or there is a distribution of various kinds of active sites, each species leading to its proper molecular weight distribution. The first explanation (diffusion control) is not in line with a first order versus monomer concentration often observed by many authors. In addition, its possible effect on the molecular weight distribution has been calculated by Ray et al (59) as limited to a Mw/Mn ratio, not larger than 11, which is far from the experimental value. Finally, recent discussion by Laurence (60) concluded that, in gas phase polymerization with catalyst fragmentation at the very beginning of the polymerization, which are the conditions observed in the Phillips system, except for the presence of a volatil diluent, there is no significant monomer diffusion effect. So, it must be concluded that the broadness of the molecular weight distribution must be explained mainly by a large distribution of active sites with different properties.The direct correlation between the nature of the active sites and the control of the molecular weight is supported by the results reported by two different teams (35, 36, 61, 62) about the effect of the introduction of titanium in the catalyst and the value of the melt index and also the broadness of the distribution ; in both cases, adequat introduction of titanium caused enhancement of both the catalytic activity and of the melt index value, while Pullukat (61) reported a greater Mw/Mn value.

This conclusion is at variance to several authors who claimed that the active sites are isolated chromium atoms with a given oxidation number either 2 or 3 (30, 31). However, it might be possible to consider isolated chromium atoms with differentiated oxidation numbers or coordination environment. It remains very difficult to conclude about the oxidation number of the active sites. The arguments given recently by different authors (63, 64) for Cr III active sites are chiefly based on studies concerning precursors of the actual catalysts, prior to the contact with ethylene ; on the other hand, Mc Daniel (65) using XPS experiments, gave strong evidence for most of the chromium being divalent after the initiation of the polymerization ; it must be recalled that the last step of the activation of the catalysts is always carried out by the monomer itself. Other changes in the environment of the active chromium atoms are concerned with their links with the support. Clearly, when silanol groups are present, they are the preferential loci for attack by chromic acid to give mono- di- or even polychromates ; but, as shown by Mc Daniel (62), the most active catalysts are obtained when the silanol groups have been previously reacted with $TiCl_4$ and the chromium moieties introduced in a waterfree and aprotic medium. In the conventional catalysts, the chromium is expected to be linked to the silica by one (dichromate) or two (monochromates) Cr-O-Si bonds ; however, some sites (trivalent Cr) may have a further Cr-O-Si bond. In addition, coordination of the chromium with the oxygen atom of Si-O-Si surface bond is possible and has been previously suggested. When titanium is present, new possibilites arise, such as the formation of Ti-O-Cr bonds, as suggested by Pullukat (35). Spectroscopic studies of CO or NO adsorption point out the existence of different kinds of links between the chromium atom and the adsorbed gas (66, 67) : four different kinds of Cr atoms, previously reduced with CO and the supposedly divalent chromium, were observed. Two of them were exposed and were readily oxidized (at room temperature) and form dinotrosyl species ; the other two, more blocked in the catalyst, are oxidized only at high temperature and form mononitrosyl species except under high NO pressure. The differences in their energetic and kinetic behavior may be expected to give rise to different kinds of sites with respect to their ability to coordinate and insert ethylene molecules.

All the above mentioned kinds of differences (oxidation number, link with the support, coordinative properties) may be expected to exist for isolated, as well as for non isolated chromium atoms ; however, new possibilities are valid for the latter species, because now, one or more chromium atoms of various nature (oxidation number ...) may exist in the second coordination sphere of the active site. The reduction profile mentioned above, and similar to those reported by Holm and Clark (23) or Hogan (29) show that, whatever the chromium contents, a variety of chromium sites with different reducibility are present ; in addition, the relative importance of the high temperature tail shows that the more isolated chromium atoms (obviously relatively more abundant for low chromium contents) are the more difficult to be reduced ; further our experiments, in agreement with the data reported by Eden (20) and more recently by Groeneveld (68), show that the less reducible chromium atoms are not the most catalytically active ones. The most active sites most probably belong to small aggreggates of chromium oxide ; such a statement is supported by the fact that, upon introduction of titanium in the catalyst, both the catalytic activity and the aggregation degree of the chromium are enhanced.However,a more precise description of the catalytic sites is very difficult, precisely because it must exist a variety of active sites with different properties, activity and specificity.

CNRS - Laboratoire des Materiaux Organiques
BP 24 - 69390 VERNAISON (France)

REFERENCES

1 Belgium Patent - 530617 (1954)

2 A. GUYOT and J.C. DANIEL - J. Polym. Sci. -1 2989 (1963)

3 A. GUYOT, C. BLANC, J.C. DANIEL et Y. TRAMBOUZE - Compt.-Rend. -253 1795 (1961)

4 A. REVILLON and A. GUYOT - J. Chim. Phys. - 65 845 (1968)

5 A. REVILLON, A. GUYOT and Y. TRAMBOUZE - J. Chim. Phys. - 65 849 (1968)

6 A. GUYOT, J.C. DANIEL, M. DURRIEU and P. PTACK - J. Polym. Sci. - A3 1765 (1965)

7 G. VUILLAUME, A. REVILLON, R. SPITZ and A. GUYOT - J. Macromol. Sci. Chem. -A5 559 (1971)

8 R. SPITZ - J. Macromol. Sci. Chem. -A6 169 (1972)

9 G. VUILLAUME, R. SPITZ, A. REVILLON and A. GUYOT - J. Macromol. Sci. Chem. -A8 1117 (1974)

10 K. V. TOPCHIEVA et al. - Vth World European Petroleum Congress - New York 1959, Sect. 3, paper 10

11 A. GUYOT, H. CHARCOSSET, A. REVILLON and G. VUILLAUME - J. Macromol. Sci. Chem. -A2 135 (1968)

12 H. CHARCOSSET, A. REVILLON and A. GUYOT - J. Catal.- 8 326-334 (1967)

13 L. NEEL - C.R. Acad. Sci. - 2552 4075 (1961)
J. Phys. Soc. Japan 17 Suppl. B1, 676 (1962)

14 P. TURLIER, M. DURRIEU AND A. GUYOT - J. Catal. - 9 295 (1967)

15 A. CLARK, J.N. FINCH and B.H. ASHE - Proc. 3rd Int.Cong. on Catalysis - Amsterdam 1964 - North Holland Publ. Co. - p. 1010 (1965)

16 R. SPITZ, G. VUILLAUME, A. REVILLON and A. GUYOT - J. Macromol. Sci. Chem. -A6 153 (1972)

17 G. VUILLAUME, R. SPITZ, A. REVILLON, H. CHARCOSSET, P. TURLIER and A. GUYOT - J. Catal. - 21 159 (1971)

18 P.B. AYSCOUGH, C. EDEN and H. STEINER - J. Catal. - 4 268 (1965)

19 V.B. KAZANSKI and J. TURKEVICH - J. Catal. 8 231 (1967)

20 C. EDEN, H. FEILCHENFELD and Y. HAAS - J. Catal. - 9 367 (1967)

21 J. HABESHAW and T. HILL - Proc. 3rd Int. Cong. Catal.Amsterdam 1964 - North Holland Publ. Co, p. 975 (1965)

22 T. HILL - Discuss. Faraday Soc. - 41 314 (1966)

23 V.C.F. HOLM and A. CLARK - J. Catal. - 11 305 (1968)

24 R. SPITZ, A. REVILLON and A. GUYOT - J. Catal. - 35 335 (1974)

25 H.L. KRAUSS and H. STACH - Inorg. Nucl. Chem. Lett.- 4 393 (1968)

26 R. SPITZ - J. Catal. - 35 345 (1974)

27 A. ELLISON and K.S.W. SING - Discuss. Faraday Soc. - 41 315 (1966)

28 R. SPITZ, A. REVILLON and A. GUYOT - J. Macromol. Sci. Chem. -A8 1129 (1974)

29 J.P. HOGAN - J. Polym. Sci. -A1 2637 (1970)

30 H.L. KRAUSS - Proc. 5th Int. Cong. Catalysis Miami 1972 - Vol. 8 North Holland Amsterdam 1973, p. 207

31 M.P. Mc DANIEL - J. Catal. 76 17 29 (1982)

32 A. ZECCHINA, E. GARONE, G. GHIOTTI, C. MORTERRA and E. BORELLO - J. Phys. Chem. 79 966 (1975)

33 J.R. PLADZIEWICE and J.H. ESPENSON - Inorg. Chem. - 10 634 (1971)

34 D.D. ELEY, C.H. ROCHESTER and M.S. CURRELL - Proc. Roy. Soc. Londond A329 361 (1972)

35 T.J. PULLUKAT, R.E. HOFF and M. SHIDA - J. Polym. Chem. Ed. 18 2857 (1980)

36 M.P. Mc DANIEL, M.B. WELCH and M.J. DREILING J. Catal. 82 118 (1983)

37 J.P. HOGAN, R.L. BANKS - US Patent 2825721 (1953)

38 R. SPITZ, B. FLORIN and A. GUYOT - Europ. Polym. J. 15 441 (1979)

39 B. FLORIN, M.F. LLAURO, R. SPITZ and Q.T. PHAM Europ. Polym. J. 15 277 (1979)

40 B. FLORIN, R. SPITZ, A. DOUILLARD, A. GUYOT, R.F. BOYER, L. RICHARDS DENNY and P.L. KUMLER Europ. Polym. J. 16 1079 (1900)

41 G. MABILON and R. SPITZ - Europ. Polym. J. (in press)

42 Y.I. YERMAKOV, B.N. KUZNETSOV and V.A. ZAKHAROV "Catalysis by supported complexes" - Vol. 8 of the series "Studies in surface science and catalysis" - Elsevier Publ. Co. Amsterdam 1981-p. 159

43 L.L. VAN REIJEN and P. COSSEE - Discuss. Faraday Soc. - 41 277 (1966)

44 Y.I. YERMAKOV and V.A. ZAKHAROV - Adv. Catal. - 24 173 (1975)

45 G. WILKE, B. BOGDANOVIC, P. HAROT, P. HEIMBACH, W. KEIM, M. KRONER, W. OBERKIRCH, K. TANAKA, E. STENRUCKE, D. WALTER and H. ZIMMERMANN - Angew. Chem. 78 157 (1966)

46 V.A. ZAKAHROV - Kinet. i Katal.21 892 (1980)

47 B. REBENSTORF and R. LARSSON - J. Mol. Catal.11 247 (1981)

48 B. REBENSTORF - Z. Anorg. Allg. Chem. 513 103 (1984)

49 H.L. KRAUSS and E. HUMS - Z. Naturforsch. 34b 1628 (1979)

50 J.L. HERISSON and Y. CHAUVIN - Makromol. Chem. 141 161 (1970)

51 M.L.H. GREEN - Pure and Applied Chem. 50 27 (1978)

52 A. ZAMBELLI, P. LOCATELLI, M.L. SACCHI, E. RIGAMONTI, Macromolecules 13 798 (1980)

53 H.W. TURNER and R.R. SCHROCK - J. Am. Chem. Soc. 104 2331 (1982)

54 H.L. KRAUSS anfd H. SCHMIDT - Reaction Kinetics and Catal. Letters 3 83 (1975)

55 R. AUMANN and E.O. FISHER - Angew. Chem. Int. Ed. 6181 (1967)

56 G. MABILON - Thesis Lyon 1982

57 H.L. KRAUSS - Proceedings 28th IUPAC Macromolecular Symposium Amherst July 1982, p. 238

58 C. GROENEVELD, P.P. WITGEN, H.P. SWINNEN, A. WERNSEN and G.C. SCHUIT - J. Catal. 83 346 (1983)

59 T.W. TAYLOR, K.Y. CHOI, H. YUAN and W.H. RAY in R.P. Quirk "Transition Metal Catalyzed Polymerization" MMI Press Symposium Series - Vol. 4, p 191 Harwoord Acad. Publ. (1983)

60 R.L. LAURENCE and M.G. CHIOVETTA In "Polymer Reaction Engineering" edited by K.H. Reichert and . Geiseler Hansen Publ. Munich 1983, p 72/111

61 T.J. PULLUKAT, R.E. HOFF and M. SHIDA in R.P. Quirk "Transition Metal Catalyzed Polymerization" MMI Press Symposium Series - Vol. 4, p 697 Harwoord Acad. Publ. (1983)

62 M.P. Mc Daniel - J. Catal. 82 98 110 (1983)

63 D.D. BECK and J.H. LUNSFORD - J. Catal. 68 121 (1981)

64 L.K. PRZHEVAL'SKAYA, V.A. SHVETS,V.B. KAZANSKY, G.L. HALLER and G.L. MORGAN - Kinet. i Katal. 25 236 (1984)

65 M.P. Mc DANIEL in R.P. QUIRCK "Transition metal catalyzed polymerization" - op. cited (59, 61)_ p. 713

66 B. FUBINI, G. GHIOTTI, L. STRADELLA, E. GARRONE and C. MORTERRA - J. Catal. 66 200 (1980)

67 G. GHIOTTI, E. GARRONE, G. DELLAGATTA, B. FUBINI and E. GIAMELLO - J. Catal. 80 249 (1983)

68 C. GROENEVELD, P.P. WITTGEN, J.P. LAVRIJSEN and G.C. SCHUIT - J. Catal. 82 77 (1983)

Ronald C. Kowalski

CONTROLLED RHEOLOGY PP RESINS

The development of controlled rheology (CR) products and of the process for making them began with recognition of the important negative impact of polypropylene's very high melt elasticity on the processor's high-speed operations.

The second necessary concept, quite well accepted today but not 20-25 years ago, was that this very high elasticity could be reduced by narrowing in a specific way the typical broad molecular weight distribution produced by Ziegler-Natta catalysis. The specific narrowing desired was removal of the high molecular weight tail of the typical distribution. Some experimental work by Rudd[1] on polystyrene and later by Kowalski[2] on polypropylene had suggested that this would be possible.

The third idea was not a new one, and could be understood on a statistical basis, that a random degradation process could produce the distribution narrowing proposed by the second concept.

The most strongly argued of the three among the experts at the time was the first--that its very high melt elasticity was the key to many of polypropylene's high-speed processing applications. This postulate remained unaccepted even after the development was complete, both within the Company and outside (discussion by some renowned rheologists following the author's paper[3] at the Fifth International Congress of Rheology). The values of negative criticism in the formative period of an innovation (as distinct from only an invention) are powerful ones, and the effects can be positive as well as negative. The author is planning a manuscript on the subject of industrial innovation in which those values are discussed in depth.

Several forward-thinking managers at Exxon Chemical played important roles at three key points in the development of CR technology, and their specific contributions are part of its history. I will point to these as they arise.

The root of the first research at Exxon Chemical in

R. B. Seymour and T. Cheng (eds.), History of Polyolefins, 307–318.

polypropylene melt rheology is one of those points. Many of you will know the name of Dr. Francis P. Baldwin as an outstanding polymer research chemist--winner of the Rubber Division's Goodyear Medal[4]. Some of his work in elastomers is in fact described elsewhere in this symposium[5]. You may not be aware that at one time he was a reluctant Section Head of the PP Polymer Characterization Group at Exxon Chemical. This brief, failed attempt to develop him into a manager was to my career good fortune because I was assigned to work in his group for the summer of 1958 when I was a graduate student at PINY (then Polytechnic Institute of Brooklyn). Baldwin combined a need of his with some experience in my background to create what grew years later into an extensive research program in melt rheology. I had been interested in it for four years at that time. First, as a result of another summer job with a small plastics extrusion company, developing a new hot melt pressure-sensitive adhesive. Then as a senior-year student of Prof. A. X. Schmidt[6], studying polymer theory (rheology emphasized) and as a graduate student at Brooklyn Poly, minoring in the renowned Polymer Institute.

Baldwin asked me to look at an unused capillary rheometer designed by Ed Severs[7] which had been gathering dust in the basement to "see if it was good for anything". That began a series of summer projects with that instrument and others[8] which grew into a group as large as five full-time professionals studying various aspects of PP's limitations in rheology. A key figure in forming that group was Dr. W. J. G. McCulloch, who saw that the early work had value and was likely to help his creation of a strong PP marketing technical service organization. Studies done in that organization in the early '60's by this author and others began to correlate some of the field experience for polypropylene with variations in its melt viscosity and, later, melt elasticity. After the first summer's work with Baldy, I proposed as my Ph.D. Thesis subject at Brooklyn Poly a study of Polymer Melt Elasticity[2], concentrating on PP.

For that work a new rheometer[8] was developed as well as a technique for measuring die swell as a manifestation of melt elasticity. The swell measured was shown to be what remained from a retarded elastic recovery of strain imposed in the capillary shear field and in the reservoir entrance to the capillary, Figures 1 and 2. Particularly important for the later CR work were the data showing that the time

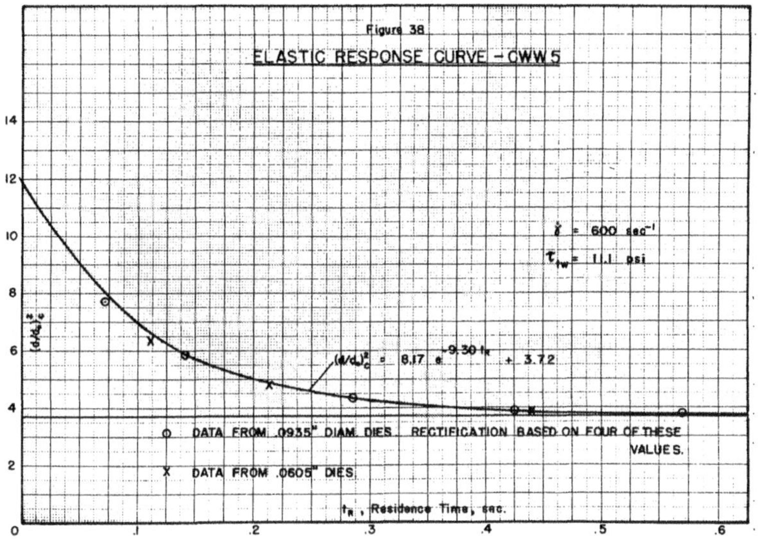

Figure 1.

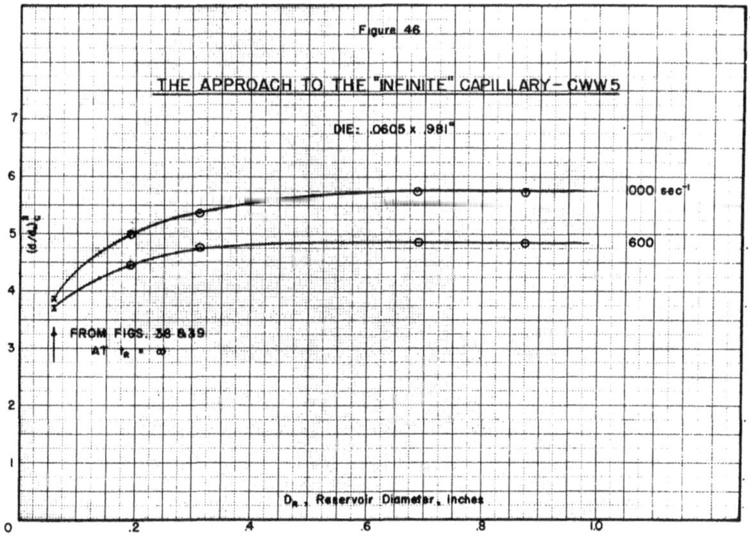

Figure 2.

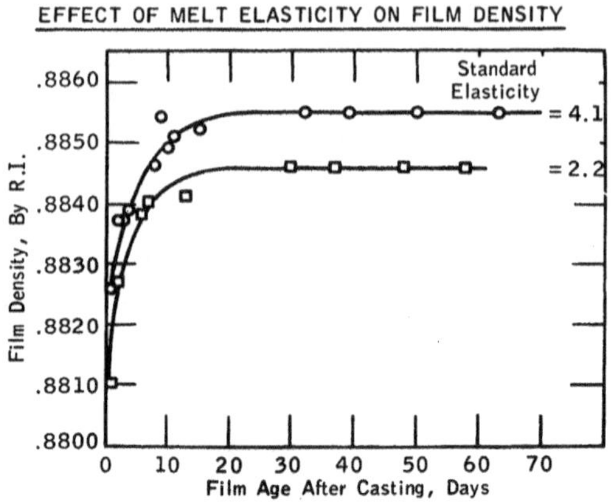

Figure 3.

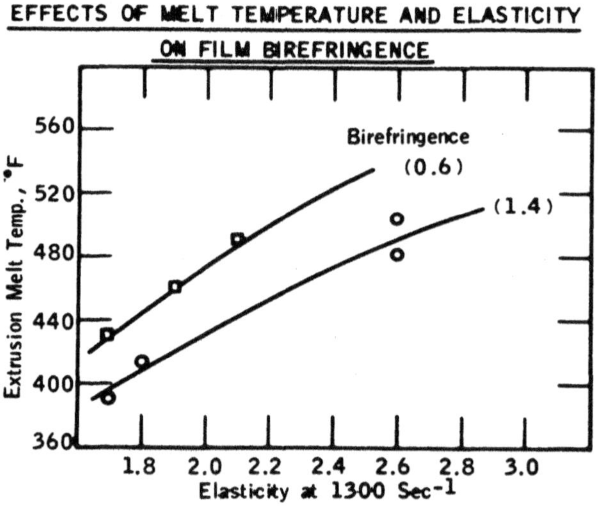

Figure 4.

scale for recovery was in tenths of a second. With this came the recognition that especially in high-speed processing (film, fiber, etc.), much of this high strain would be frozen into the product, creating stresses which were not readily controllable and which would affect the product's physical properties. Much later in 1968, when low swell resins were available from the CR process, some of these effects were shown[3], Figures 3, 4, 5, 6.

Returning to the mid-'60's, some very important support from two other parts of the organization was received when this author made a proposal that Exxon Chemical found a research project to study extrusion degradation as a means to narrow the molecular weight distribution, and hopefully to reduce the elasticity and its postulated debits. Dr. A. T. Watson, manager of the PP research laboratory in Baytown, Texas, accepted the proposal and through his unique blend of encouragement, challenge, technical criticism and personal involvement, provided management guidance through a period when the value of the project was not universally accepted. The late Dr. C. L. Umholtz of the Manufacturing Department was also an early supporter and understood the potential of the connection of the concepts of extrusion degradation, MWD, melt elasticity and processing. His support was not only influential in providing a "customer" for the proposed technology, but also in the first year of the work, in opening his plant for commercial testing of the process as it was being developed in the research laboratories.

The academic research[2] had shown the value of a shorthand technique for characterizing and pictorializing the relations between standardized viscosity and elasticity measurements and molecular weight and MWD, the "Viscoelastic Grid", Figure 7. The research group, consisting in 1965 of Dr. J. W. Harrison, J. C. Staton, J. P. Keller and the author, showed quantitatively that this was a valid concept, Figures 8 and 9, and that it could be used for many purposes, including plotting the course of the extrusion degradation process being developed by that group, Figure 10. The technology of the process is described in a series of patents[9,10,11] which became the basis for the first commercial process. For that process we coined the term "Controlled Rheology", widely used in the industry today[12,13].

The process was found to be highly controllable in terms of the temperature profile in the extruder, the patents showing that a "reverse temperature profile" was

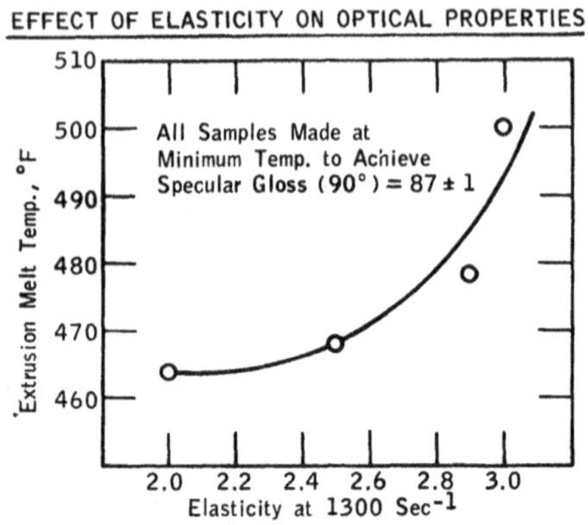

Figure 5.

EFFECT OF ELASTICITY AND TEMPERATURE ON PHYSICAL PROPERTIES AND BIREFRINGENCE OF FILMS

Standard Elasticity of Melt	Impact Strength, Ft-Lb/Mil.	Secant Modulus ($Lb/In.^2$)	Birefringence, x 10^3	Extrusion Melt Temp, °F
1.8	>0.8	91,000	1.58	414
1.9	>0.8	94,000	0.55	460
2.1	>0.8	101,000	0.53	485
2.6	>0.8	99,000	1.51	480
2.6	>0.8	103,000	1.41	507
7.4	0.3	122,000	2.09	527

Figure 6.

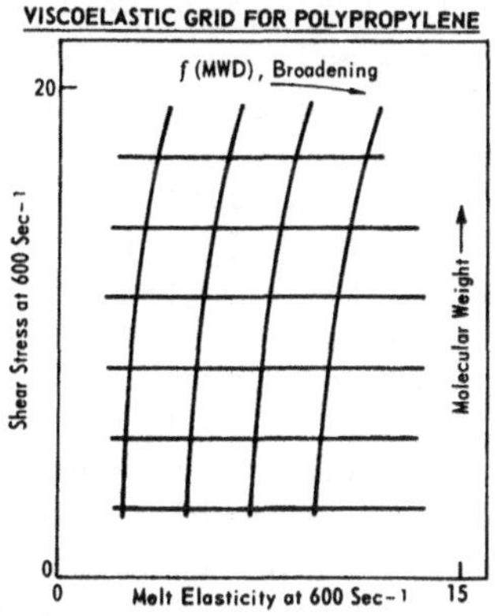

Figure 7.

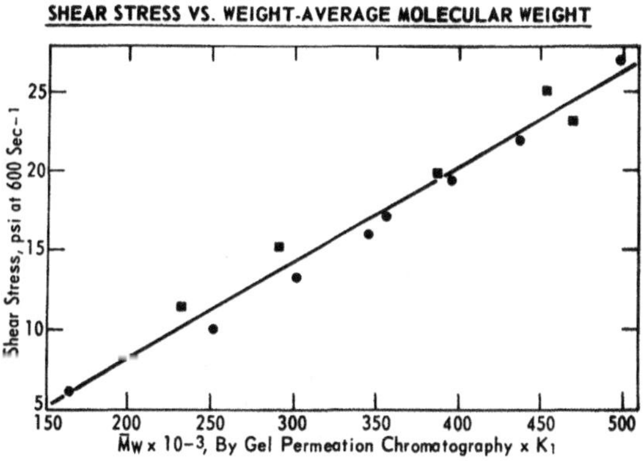

Figure 8.

advantageous, wherein the highest and controlling temperature was imposed as early as possible in the extruder where oxygen was still available to serve as a free-radical initiator. This came as a result of the group's finding that the process was not governed by mechanical shear, but by an oxygen-induced, free-radical chain scission chemical reaction. [This recognition became a precursor to research later into other chemical reactions which could be done in extruders[14,15]. This author will present a paper next month at an ACS Topical Workshop on that subject.]

Development of a short-residence-time, low-inventory process which effects a dramatic property change on the product creates both a problem and an opportunity. Large-scale blending or process surge damping is not available in the reactor, and post-blending may not be feasible, even considering its extra cost. The research group recognized this problem and that its solution, a real-time on-line analyzer, would solve it and provide the opportunity for a process in tight control, responding quickly to process drifts. The opportunity presented is for a process providing minimum off-spec, immediate diversion of any off-spec, no post-blending requirements and therefore maximum product uniformity. For these reasons the group developed and built an on-line, continuous rheometer, in parallel with the process and product research program. The property measured was the standard Shear Stress coordinate of the Viscoelastic Grid. If the coordinates of the feedstock were premeasured, a target of rheometer Shear Stress could be defined which fixed both the viscosity and the elasticity of the product. The continuous rheometer measurement was done on pellets immediately after they were cut upon exit from the CR extruder-reactor.

When the technology was available for first commercial trials at the end of 1965, two units of that instrument were also available, Figure 11, and were used to make the first commercial-sized quantities.

With a nearly instantaneous electronic signal of the product variable available, the operator could manually respond to the first sign of process drift by adjusting the reactor temperature profile, most quickly by die pressure valve adjustment. Though it was very early in the era of computer process control, the temptation to use the electronic product signal to automatically close the control loop was irresistible. The author, together with J. Beauxis, a very resourceful electronics engineer, developed,

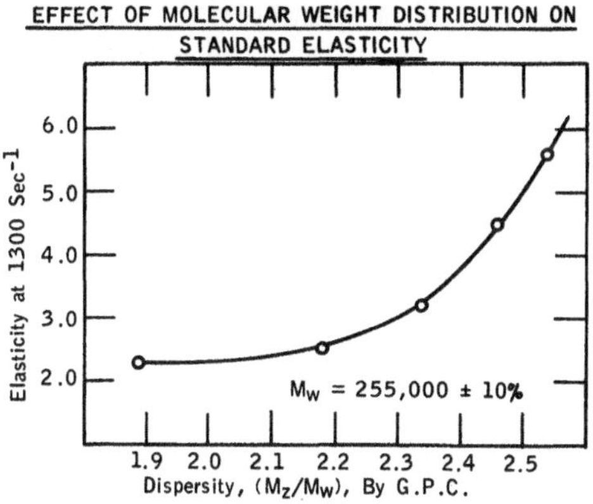

Figure 9.

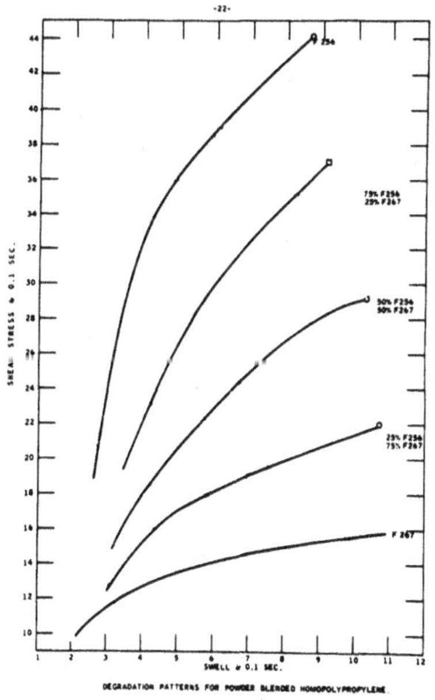

Figure 10.

built and installed from simple parts (no microprocessors, hardware, software packages available then), an analog computer control system[16]. It was demonstrated that this computer even in its first version did a better job of controlling the process than the CR guru himself! One of the reasons was that we built into it a fixed delay period of 12 minutes when no process changes could be made. No human operator could match that patience and discretion.

The above accomplishments having been made, the innovation was not yet complete. Invention, research, process and product development, even demonstration of commercial operability taken together create no new value for the world. Commercial use and return on investment are also required. This last hurdle can often be the most difficult for truly new technologies, because those responsible for the P/L require convincing that a real market will be there in the future, while no customers are asking for the yet unavailable commercial product with yet unconfirmed (in hard business terms) dream properties. The dreamer is often the least convincing proponent to the business manager, the marketplace the most convincing. In this instance we were able to "arrange" for a partially captive customer to convince our marketers via a strong financial message that the dream was real. Although 20 years have passed, the details of this "arrangement", though fascinating, will have to wait a bit longer before they are fully exposed. But the innovation was in fact completed and, to the best of my knowledge, about seven years elapsed before any other competitor marketed commercial CR products.

Gifford Pinchot III[17] has written recently on the subject of innovation in large companies, making the point that it is indeed a feasible opportunity. The CR example, as well as some others I can mention, supports that argument. The innovation process is not without difficulties, more perhaps and different ones than in small organizations. The impact of people and their individual peculiarities is at least as important as the technology or economics in determining whether the innovation is made. This author will have more to say in other contexts about the psychology of industrial innovations.

Exxon Chemical Company
P. O. Box 45
Linden, New Jersey 07036

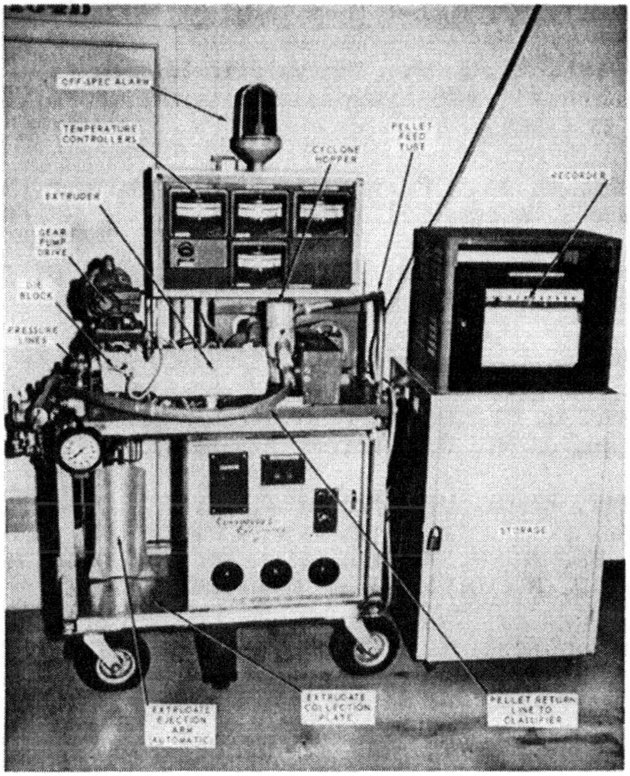

Figure 11.

REFERENCES

1. Rudd, J. F., J. Poly. Sci., 44, 459 (1960).

2. Kowalski, R. C. 'Elastic Behavior in Molten Viscoelastic Polymeric Materials', Ph.D. Thesis, Polytech. Inst., Brooklyn, (1963).

3. Kowalski, R.C., Fifth International Congréss of Rheology, Kyoto, Japan (1968).

4. Baldwin, F. P., Rubber Chem Tech. 52, 677 (1979).

5. Tornqvist, E. G. M., 'Polyolefin Elastomers - 50 Years of Progress', ACS Symposium History of Polyolefins, April 29 (1985).

6. Schmidt, A. X., 'Principles of High Polymer Theory and Practice', McGraw-Hill (1948).

7. Severs, E. T., Ph.D. Thesis, University of Delaware (1950).

8. Kowalski R. C., U.S. Patent 3,279,240, 10/18/66.

9. Staton, J. C., Keller, J. P., Kowalski, R. C. and Harrison, J. W., U.S. Patent 3,551,943, 1/5/71.

10. Kowalski, R. C., U.S. Patent 3,563,972, 2/16/71.

11. Kowalski, R. C. Harrison, J. W., Staton, J. C. and Keller, J. P., U.S. Patent 3,608,001, 9/21/71.

12. Dziemianowicz, T. S. and Cox, W. W., 'Crystallization Behavior of Controlled Rheology Polypropylene', SPE ANTEC, 4/29/85.

13. Brockschmidt, A., Plastics Technology, 67, March, 1982.

14. Steinkamp, R. and Grail, T., U.S. Patent 3,862,265, 1/21/75.

15. Newman, N. and Kowalski, R. C., U.S. Patent 4,384,072, 5/17/83 et seq.

16. Beauxis, J. and Kowalski, R. C. , U.S. Patent 3,593,011, 7/13/71.

17. Pinchot, G., 'Intrapreneuring', Harper and Row (1985).

Raymond B. Seymour

CONCLUSIONS

As stated in the preface by the eminent polymer scientist, Herman F. Mark, the word polyethylene is over 120 years old. What is now known as high density polyethylene (HDPE) was made by Von Pechmann and others around the beginning of the 20th century and again by another eminent polymer scientist, Carl S. Marvel in 1930.

While duPont failed to visualize the importance of HDPE, ICI did recognize the importance of low density polyethylene (LDPE). The story of the development, based on about 0.5 g of this new polymer to an annual production of over 700,000 tons, is one of unequalled scientific adventure.

The story becomes more intriguing when one recognizes that the unheard of pressure of 2,000 atms or 200 MPa was used to produce this new material commercially at the beginning of World War II. LDPE, which was unknown prior to World War II, was essential for insulation for the newly discovered Radio Direction and Range Finding (Radar). The Royal Air Force was able to win the Battle of Britain because of the availability of LDPE for Radar insulators. This highly branced LDPE was produced in tubular reactors as long as 1500 m and also in autoclaves at pressures as high as 300 MPa at temperatures as high as 250°C.

The linear polymer (HDPE) was produced by M. Fischer in the laboratory of Badische Aniline Und Soda Fabrik in 1943 but was not patented until 1953. Karl Ziegler used a catalyst similar to that used by Fischer but unlike Fischer, Ziegler followed his discovery through to commercial production.

Ziegler served as his own patent attorney and did not include propylene in his claims. Hence, Giulio Natta was able to use the Ziegler Catalyst (now called the Ziegler-Natta Catalyst) to produce isotactic polypropylene. Many others produed polyolefins in their laboratories but, as stated by Dr. M. McMillan, at the History of Polyolefins Symposium in 1985, these investigators were not permitted

R. B. Seymour and T. Cheng (eds.), History of Polyolefins, 319–321.

or encouraged to publish their discoveries in the scientific literature.

In the late 1970's, chemists, at Union Carbide Company and Dow Chemical Company, produced a linear low density polyethylene (LLDPE) by copolymerizing ethylene and C_4-C_8 α-olefins. The low pressure slurry process used butene-1, pentene-1 and hexene-1 as comonomers. Octene-1, 4 methylpentene-1 and hexene-1 are used as comonomers in the solution polymerization process. Highly active Ziegler-Natta type catalysts, which are used, are left in the LLDPE.

While a few participants in the symposium did not submit chapters for the book, it may be in order to summarize some of their reports presented at the national American Chemical Society meeting at Miami Beach, FL.

According to Dr. R.S. Stein, early investigators described the orientation of crystalline polyolefins, using the Kratky "floating rod" model. More recently, the parameters essential for modeling mechanical behavior of these crystalline products have been developed by microscopy, diffraction, scattering and rheo-optical techniques complemented by small angle light and neutron scattering methods. These techniques have permitted the description of changes in chain topology, crystalline, amorphous and lamellar orientations and spherulite deformation accompanying sample deformation.

According to Dr. William (Bill) Bailey, the discovery of crystalline polypropylene by J.P. Hogan and R.L. Banks triggered one of the longest and most complicated patent interference proceedings of all time. After an interference suit by Standard Oil of Indiana, Montedison, duPont and Phillips, U.S. Patent No. 4,376,852 was issued to Phillips on March 15 1981.

Hogan and Banks did not use the words "crystalline" in their patent application in 1953 but they did show a melting point transition in their cooling curve which was consistent with crystallinity. Natta's proof of structure of isotactic polypropylene was reported in 1954. The solid polymer product produced earlier by Standard Oil of Indiana was believed to be a copolymer of propylene, ethylene and 2-butene and not solid crystalline polypropylene. The basis for this conclusion was the fact that propylene disproportionates to ethylene and 2-butene in the presence of the molybdenum catalyst used by Standard Oil of Indiana.

According to Dr. P. Pino, a senior member of Dr. Natta's staff, Dr. Hermann Staudinger was aware of the existence of stero regular polystyrene in the 1920's and in 1944, Dr. M. Huggins used this concept to explain some of the properties of solutions of polystyrene. However, the breakthrough occurred in 1954, with his first synthesis of polypropylene, using the Ziegler catalyst, Natta recognized the sterospecific character of the polymer. Thus, on March 11, 1951, with no evidence of emotion, Dr. Natta made the following statement in his notebook, "Today, we made polypropylene,"

As stated in Chapter 9 by Dr. Frederick J. Karol, the development of new catalysts has led to a polyethylene revolution. Unlike military revolutions, this revolution will not be terminated quickly. Much history on polyolefins will be recorded in the future. However, it is important to record the progress of the past half century so that much greater progress can be made in the future.

INDEX

GPSR Compliance

The European Union's (EU) General Product Safety Regulation (GPSR) is a set of rules that requires consumer products to be safe and our obligations to ensure this.

If you have any concerns about our products, you can contact us on ProductSafety@springernature.com

In case Publisher is established outside the EU, the EU authorized representative is:

Springer Nature Customer Service Center GmbH
Europaplatz 3
69115 Heidelberg, Germany

Batch number: 10761599

Printed by Printforce, the Netherlands